PRAISE FOR *REINVENTING THE PRODUCT*

'Eric Schaeffer and David Sovie bring some great new insights into the product arena of the future that have broad implications. With grounded skill and enthusiasm, *Reinventing the Product* makes a stringent case for companies to rethink their product strategy, their product roadmap and their digital capabilities.'
Patrick Koller, CEO, Faurecia

'Powerful factors – such as the rapid rise of cloud computing, high-speed networks and AI – are converging. They require all product companies to fundamentally transform their products and their company. Dave Sovie and Eric Schaeffer bring fresh thinking and inspiring practical advice for successfully managing this digital transformation and creating value.'
Marco Argenti, Vice President Technology, Amazon Web Services

'Thoroughly researched and full of innovative insights about AI, platforms and smart products. The "Product Reinvention Quotient" provides great insights in how to think about "Product X.0" and how to develop a set of capabilities that are necessary to succeed in the future.'
Guido Jouret, Chief Digital Officer, ABB

'The nature of product innovation is fundamentally changing. *Reinventing the Product* shows how to combine hardware, software and business model innovation in an agile manner to meet fast-changing needs in a world of smart connected devices. And it provides compelling and inspiring case studies and examples that help to understand how your company can find its way that fits best.'
Yoon Lee, PhD, Senior Vice President and Division Head: Content and Services, Product Innovation Team, Samsung Electronics America

'Eric Schaeffer did it again. After his spearheading *Industry X.0*, which has inspired our team of entrepreneurs to adapt our digital strategy creatively, he now co-authors with David Sovie a brilliant, richly detailed roadmap to the digital future for all product-making companies. No doubt: this is again a must-read.'

Zhang Ruimin, Chairman of the Board of Directors and CEO, Haier Group

'Well researched, with vivid illustrations and concrete suggestions, this valuable guide can help firms and leaders to build a new set of priorities and capabilities to succeed in a shifting, digital landscape.'

**Professor Michael G Jacobides, Sir Donald Gordon Chair
of Entrepreneurship & Innovation, London Business School**

'In a time of digitally induced seismic shifts across all fronts, *Reinventing the Product* captures the impact of this change and thoughtfully develops new value creation approaches in product development and manufacturing. A ground-breaking book.'

**Phil Jansen, Vice President and Head of Product Development, Fiat
Chrysler Automobiles**

'*Reinventing the Product* is a practical guide for harnessing IoT and AI to transform the very basis of a company's offerings. This is a must-read for industrial manufacturers looking to ensure their businesses remain relevant in the digital age.'

Tim O'Keeffe, CEO, Symmons Industries

'Digital technologies such as AI, advanced analytics, edge computing, cloud and blockchain are transforming our lives fast. Thoroughly researched, *Reinventing the Product* doesn't just describe the emergence of a fascinating new product landscape that is shifting from traditional to smart connected products, including autonomous products. It is also an inspiring call-to-action for companies so seize enormous new opportunities fast – good for them, good for users and customers, good for growth and progress of societies.'

**Professor Dr Christoph Lütge, Peter Löscher Chair of Business Ethics
and Global Governance, Technical University of Munich**

'Dave Sovie and Eric Schaeffer lay out a powerful new framework for how to evolve both the product and business strategy needed to succeed in the digital age, and they give fresh and very concrete recommendations on how to implement it pragmatically. *Reinventing the Product* should be required reading for all product company executives and their managers.'
Bill Bien, Chief Marketing Officer and Head of Strategy, Signify

'Deep analysis of major shifts product companies are facing, and a well-elaborated blueprint for their future success. Every leader of this industry should learn from this book.'
James E Heppelmann, President and CEO, PTC

'*Reinventing the Product* doesn't just look thoroughly at how the disruptive waves of digital technologies will affect product companies (including the subscription economy). Drawing on a deep analysis of five profound shifts they face and complemented with compelling case studies, Eric Schaeffer and David Sovie also provide strategic and practical how-to advice for businesses as they develop digital products. An inspirational call-to-action.'
Eric Chaniot, Chief Digital Officer, Michelin

'The swiftly emerging world of intelligent, smart connected products will reshape industries, business processes, and consumer experiences. Schaeffer and Sovie's book is an important and essential guide for business leaders, entrepreneurs and investors looking to chart the course and unlock the value of this important trend.'
Paul R Daugherty, Chief Technology & Innovation Officer, Accenture, and co-author of *Human + Machine: Reimagining Work in the Age of AI*

'A comprehensive analysis on the digitally driven big shifts product-making companies are facing, and a detailed roadmap to innovate and capture the endless opportunities in a fascinating new product world.'
Raghunath Mashelkar, National Research Professor and Chairman, Reliance Innovation Council

'In their inspiring new book Eric Schaeffer and Dave Sovie not only show the fundamental shifts product companies are exposed to in digital times; they also provide creative analytical tools and concrete how-to advice for these companies to innovate, stay profitable and grow.'
Cyril Perducat, Executive Vice President, Internet of Things & Digital Offers, Schneider Electric

'Eric Schaeffer and Dave Sovie provide a rare encompassing view on the reinvention of the product, re-imagining the current "digital transformation" trend. Original, remarkably thoughtful and with high practical relevance on every page.'
Richard Mark Soley, PhD, Chairman and CEO, Object Management Group; Executive Director, Industrial Internet Consortium

'This is a fascinating book on how we should rethink and manage product making in disruptive times. New perspectives, fresh concepts, unexpected ideas abound. A must-read for any leader and manager in product companies.'
Pascal Daloz, Executive Vice President, Chief Financial Officer and Corporate Strategy Officer, Dassault Systèmes

Reinventing the Product

How to transform your business
and create value in the digital age

Eric Schaeffer and David Sovie

KoganPage

First published in Great Britain and the United States in 2019 by Kogan Page Limited

2nd Floor, 45 Gee Street
London
EC1V 3RS
United Kingdom

c/o Martin P Hill Consulting
122 W 27th St, 10th Floor
New York, NY 10001
USA

4737/23 Ansari Road
Daryaganj
New Delhi 110002
India

www.koganpage.com

© Accenture, 2019

All figures © Accenture unless otherwise stated.

ISBN   978 0 7494 8464 4
E-ISBN 978 0 7494 8465 1

British Library Cataloguing-in-Publication Data

A CIP record for this book is available from the British Library.

Library of Congress Cataloging-in-Publication Data

Names: Schaeffer, Eric (Managing director), author. | Sovie, David, author.
Title: Reinventing the product : how to transform your business and create
   value in the digital age / Eric Schaeffer and David Sovie.
Description: 1 Edition. | New York : Kogan Page Ltd, [2019] | Includes
   bibliographical references and index.
Identifiers: LCCN 2018054275 (print) | LCCN 2019000425 (ebook) | ISBN
   9780749484651 (ebook) | ISBN 9780749484644 (hardback : alk. paper) | ISBN
   9780749484651 (eISBN)
Subjects: LCSH: New products–Management. | New products–Technological
   innovations. | Artificial intelligence–Economic aspects.
Classification: LCC HF5415.153 (ebook) | LCC HF5415.153 .S327 2019 (print) |
   DDC 658.5/75–dc23
LC record available at https://lccn.loc.gov/2018054275

Typeset by Integra Software Services, Pondicherry
Print production managed by Jellyfish
Printed and bound by CPI Group (UK) Ltd, Croydon CR0 4YY

CONTENTS

ABOUT THE AUTHORS

Eric Schaeffer

Eric Schaeffer is a Senior Managing Director at Accenture, focused on helping industrial organizations harness connected innovation to digitally transform their businesses for growth.

Over the last 30 years at Accenture he has helped industrial clients across Europe and now globally to envisage and deliver transformation by placing product innovation at the centre of the change process. His first book, *Industry X.0: Realizing digital value in industrial sectors*, published in English, German, French, Japanese, Chinese, and soon Brazilian Portuguese and Russian, examines the six fundamental digital 'no regret' capabilities every industrial business needs as a launchpad for digitization.

Schaeffer leads Accenture's Products Industry X.0 practice, bringing together innovation, engineering, product development, manufacturing, digital operations, and product service optimization to help industry clients master the opportunities of the Industrial Internet of Things. He is also the Global Lead for Automotive, Industrial Equipment, Infrastructure and Transportation, helping these companies to digitally reinvent their business.

Schaeffer's background is in engineering. He studied at the École Supérieure d'Électricité, and is based in Paris, where he lives with his family.

David Sovie

David Sovie is a Senior Managing Director at Accenture and the Global Lead for High Tech Industry. He focuses on shaping and executing large-scale digital transformation and business reinvention programmes for leading technology companies globally.

Sovie also leads the Industry X.0 practice for Accenture's Communications, Media and Technology group, which provides digital transformation services across the product innovation, engineering, manufacturing and product support business functions. The Industry X.0 practice helps clients reinvent their product and service offerings and create new revenue streams by leveraging new engineering approaches and digital technologies such as AI, advanced analytics, blockchain and augmented reality.

Sovie studied Electrical Engineering at Rensselaer Polytechnic Institute and went on to receive his MBA from Harvard Business School.

He lives in Tokyo with his wife.

FOREWORD

'When the wind of change rises, some people build walls while others erect windmills.' This is an old Chinese saying, but it is very relevant in the era of innovation that we are living in today.

There is no doubt about the rising wind: a new era of smart-connected products is in the offing; a fascinating period with lots of opportunities for businesses. But still we see many businesses staying wedded to accustomed ways of making and using industrial products, missing out on the massive potential that the fast-advancing digital technologies offer once they are put to use within products.

This important book, *Reinventing the Product: How to transform your business and create value in the digital age*, pushes back against traditional modes of thinking and operating in the product-making world. It draws business leaders' attention to the urgency of embracing and shaping the emerging global smart-connected product landscape: very quickly, we'll see large parts of this novel product category evolving from basic interactivity to very advanced intelligence, becoming 'living' products essential to businesses and consumers in daily life.

Two industry leaders and shapers, Eric Schaeffer and David Sovie, have joined forces to write the book. Their combined experience across traditional industrial sectors, along with the software and technology sectors, is a reflection on the blueprint required for smart connected products.

Schaeffer's precursor publication, *Industry X.0*, described the fundamental shift industrial product-making is undergoing as digital technology and software transforms products and business processes.

This book now offers the logical follow-up, and highlights the endgame of these trends: smart-connected intelligent products and the unprecedented user experience they deliver. And it shows how this will shape the future for businesses and consumers, and be a powerful value creator.

The future is arriving quickly, which is why this book is timely. Every day we witness seeing new smart-connected products emerge. One of my own professional focuses within this wide topic is the ever-closer relationship between humans and machines, and the enormous potential this combination delivers. Digital technology and artificial intelligence are taking centre stage as the main forces for creating the assistant and 'cobot' technologies relevant to smart-connected products in so many industrial sectors from industrial equipment to automotive to home appliances. But this is just part of the enormous overall potential of smart-connected products.

Schaeffer and Sovie trace the irreversible shifts that products are undergoing to embody intelligence, deliver rich user experiences and extend into value-creating ecosystems and platforms. They make this approachable by providing practical road maps and blueprints for building the necessary business capabilities to enter this new product world.

Reinventing the Product is an outstanding *vade mecum* for any business leader, industrialist, entrepreneur or investor. We've only seen a glimpse of the true economic potential of the world of smart-connected products. However, the wind of change has clearly risen. And this book helps product-making businesses to put up as many windmills as possible to make the most of it.

Paul R Daugherty
Chief Technology & Innovation Officer at Accenture
Co-author of Human + Machine:
Reimagining work in the age of AI
January 2019

PREFACE

How dramatically will day-to-day products change in the near future – say, a car, a light bulb, a watch, a printer, a refrigerator or industrial products like a mining truck, an industrial welding machine or medical imaging equipment? To an astonishing extent, is our rough-and-ready answer. Rapid advances in software and digital technologies will make it happen, and happen much faster than most people realize today. Arguing the detailed 'what', 'why' and 'how' is the focus of this book.

To ask the entry question would have sounded downright strange not too long ago. The products mentioned were constantly being innovated and were showing advanced sophistication. However, they were still largely passive and uncommunicative electro-mechanical devices. They served their purpose and propositions well, once their users had decided to make the most of their features. Luxury cars had air conditioning and could accelerate quickly; more budget-priced vehicles couldn't. Screwdrivers were affordable but slow and blister-prone, while power screwdrivers allowed for fast work with no injuries.

For a long time, it was a matter of preference for different product features and functions that drove us to buy a specific product. And this led their makers to cater for us by launching products in great numbers with a 'one-size-fits-many' market perspective.

However, the world is rotating rapidly towards the age of digital. Those cars, watches, refrigerators, printers and mining trucks are all well on their way to adopting rather active, fluid characteristics, steered by intelligent minds. This is why we so strongly advocate in this book that products should become smart connected comrades, co-workers and colleagues supporting everybody with their adaptable services and functionalities – in private and business life, in the office and on the shop floor.

Enabled by more and more software and artificial intelligence, a new product generation is developing on-board brains that make them cooperative, configurable and eventually even autonomous actors. This amounts to a huge change in product making and product usage.

Product making has undergone several quantum leaps in innovation. When humans discovered the productive functionality of the wheel 3,000 years ago, most likely a sense of overwhelming practicality had all of a sudden transfixed people's minds with visions about being very mobile and efficient in moving heavy loads. The invention of electricity and electric motors about 150 years ago created a whole new wave of mechanical products like light bulbs, automobiles and refrigerators. The creation of the microprocessor and personal computing in the 1980s created an initial wave of smart devices like printers, game consoles and medical imaging equipment.

This time around, as we stand at the watershed of an entirely new product age, we feel the same: a sense of limitlessness is erupting around how smart connected products, more and more enabled by artificial intelligence, can be used in ways nobody has ever thought about before. These new products are becoming so versatile, minding, obedient, trustful and helpful that they truly turn into our 'brothers in mind'.

What an exciting new world. What an exciting new product world, in which products have to be thought about in radical new ways and reinvented permanently to meet fast-changing customer needs.

In 2017, Eric published his spearheading book, *Industry X.0* – very soon published in seven languages such as English, German, Japanese, Chinese, French, Brazilian Portuguese and Russian – which analyses the broad trends towards digital technology and software in industrial sectors and shows new paths of realizing value. *Industry X.0* continues to be successful. But in the ongoing exchange with our clients we – Eric and Dave, co-authoring this new book – understood that it needs some more thinking, specifically on the new world of products that is underway. Thus, *Reinventing the Product: How to transform your business and create value in the digital age* builds on its predecessor but changes the perspective as it gives a comprehensive update with a cutting-edge focus on products and the digitally induced major shifts that are propelling them to become smart and connected.

We were able to pinpoint massive evidence about what we theoretically assert in real-life use cases. And we discovered entrepreneurs and businesses around the world who started to set the ball rolling of the smart connected products that are beginning to rule the globe. Thus, in this sense, we describe a rising new product-making world, which we may well call the age of 'Product X.0', part of the fast-developing 'Industry X.0' that frames this shift.

We invited thought leaders – pioneering clients, business practitioners and professors based in the United States, the United Kingdom, France, Germany, Italy, China, Japan and Korea – to kick the tyres on our ideas on intelligent products. For example, Eric had the honour of sitting down with Zhang Ruimin, Chairman and CEO of Haier Group, the world's largest maker of domestic appliances; he also worked with Patrick Koller and his team of Faurecia, one of the world's leading automotive suppliers. David drew on his high-profile network spanning the big names of the tech sector to interview executives at Google, Amazon, Samsung and HP Inc. They and many others helped us to develop our ideas further, step by step and quickly.

And this was necessary, since the business train towards a new smart connected product category is not inching forward. It is by now speeding forward, eager to conquer the emerging markets for this novel device category that is promising so much new value for traditional industrial and software businesses.

May we say that this book doesn't stop at assessing the 'what' and 'why'. It also strongly features 'how' the world of collaborative, responsive and autonomous products is best entered by businesses and examines in detail the steps industrial businesses need to yield the greatest positive effects and value from smart connected products.

Filled with easily accessible actionable findings and suggestions, this book should be a vital resource for business leaders at all levels, C-suite and below, as well as across all functions, helping to discover, think through, adopt and implement the capabilities and roadmap needed for their enterprises to head out for the new territory of smart connected products.

ACKNOWLEDGEMENTS

This book provides concrete advice to business leaders and gives shape to an extraordinarily multifaceted topic: the emergence and rapid evolution of the smart connected product. Being two Industry Managing Directors at Accenture leading very different, yet converging, sectors, we set out to square both our views on the topic, twisted and turned intellectually with what we found, until a compelling and likewise exciting joint line of argument emerged. In this intense but inspiring process we crystallized a unique thought leadership on a most burning topic for product making.

It would have been impossible to achieve this without the manifold contributions of a large number of knowledgeable people beyond our core book team. We, the authors, were fortunate to be able to enjoy the richest intellectual stimuli from the broadest possible expert spectrum: top industry executives, corporate thinkers, our colleagues, academics and clients, based in a wide array of countries including the United States, the United Kingdom, Germany, China, France, the Netherlands, Italy, Sweden, Switzerland, Singapore, India, Brazil and Japan.

Their input has been invaluable in shaping the general themes, observations, analyses and final hypotheses expressed. To all of them we would like, at this point, to express a great 'Thank you'. The support of every one of you individually was absolutely instrumental for making this book happen. What we jointly pulled off represents vanguard industry thinking of the finest category.

First, we would like to thank the companies that agreed to directly participate in the research for this book. Special thanks go to Faurecia, Haier, Signify (formerly Philips Lighting) and Symmons for agreeing to be featured as case studies, as well as to the executives and thought leaders we interviewed at ABB, Amazon, Caterpillar, Dassault Systèmes, Google, HP, London Business School, Mindtribe, Nytec, PTC, Samsung, and Tesla.

We also wish to thank our Accenture colleagues who directly contributed to the content creation and review process, including Sam Baker, Marc Carrel Billiard, Jean Nicolas Brun, Jean Cabanes, Kimberley Clavin, Brian Doyle, Scott Ellsworth, John Giubileo, Mary Hamilton, Brian Irwin, Lisa-Cheng Jackson, Shinichiro Kohno, Alex Kass, Giuseppe La Commare, Sarat Maitin, Davide Pugliesi, Floris Provoost, Ramadurai Ramalingam, Juergen Reers, Steve Roberts, David Rush, Shugo Sohma, Philip Vann, Cedric Vatier, Maxence Tilliette and Tunc Yorulmaz.

We also say thank you to the teams from Altitude, Design Affairs, Mackevision and Mindtribe, and Pillar Technology, creative and product engineering companies that have recently joined the Accenture family. Collectively, they expanded the breadth and depth of our market insights, thought leadership and capabilities relevant to this book, and they are really making a difference.

We would also like to thank the thousands of Accenture employees who participated in our Crowdsourcing Innovation event that formed an important basis for Part Four of our book and was facilitated by Stephanie Winters McConnell and Cameron Timmis.

Special thanks must go to Pierre Nanterme, Omar Abbosh, Sander van't Noordende and Paul Daugherty for their inspiration and continuous support and leadership on Industry X.0.

We would like to especially thank the core team who participated in the creation of this book over several months: the research team led by Raghav Narsalay, based in Mumbai, with Andreas Egetenmeyer, Abhishek Gupta, Matthias Wahrendorff and Pravi Dubey; the marketing team, led by Ulf Henning, with Gemma Catchpole, Amy Oseland and Catherine Tremblay; Jens Schadendorf, Titus Kroder, and John Moseley, who brought valuable experience and exceptional knowledge with regard to the writing and publishing of this book.

A thank you also goes to Chris Cudmore, Lachean Humphreys, Susi Lowndes, Natasha Tulett and Helen Kogan from Kogan Page, the publisher of this book, for their enduring commitment and trust in our project.

Finally, and above all, we thank both our families for their support.

Eric: I would like to thank my wife, Pascale, and my three children William, Meryl and Edouard for their continuous support during the many months it took us to write this book. After its predecessor, *Industry X.0*, and with this second opus, my aim is to provide you with many keys to shape the future, your future… this is what parents are for, no?

David: I would like to thank my parents – my late father and my mother, Bonnie Sovie – for providing me with the opportunities in life that started me on the professional journey leading to this book. And a very special thank you to my amazing wife, Atsuko Watanabe. You have been a constant source of inspiration and encouragement for over 26 years, and I could never have written this book without your support.

Introduction

The advent of the smart connected product is about to change beyond recognition the ways in which products are used and how businesses conceive, make, distribute, and support products in the market.

For the last 200 years of industrial history, a product's sale marked the end of the following value chain: inputs such as raw materials and electro-mechanical components were sourced, labour force and machines were allocated to carry out manufacturing processes, and a finished product was eventually sold on a margin that reflected selling price minus input cost. Product makers were clearly delineated as producers, acting within the defined borders of their business and within a fairly simple financial calculus. The point of purchase marked the handover of responsibility for the product to the user and that was that.

This well-worn routine is about to experience a tectonic tilt towards a new product world where traditional product-making businesses will need to apply much more fluid business practices if they want to survive. With the rise of the smart connected product the linearity of value chains defined by clear beginnings and endings within siloed business organizations will be relegated to the history books. Products that are rendered digitally intelligent will instead be designed to interact with their makers, their users and other products over their whole lifespan. Value chains will turn into value circles or even multi-directional value generating systems.

New technology, new perspectives

A product ending in the hands of a user will now be far from sold and forgotten. Once it starts a smart life in the field as part of an installed base, it will be under a regime of remote contact and permanent data exchange with its manufacturers, who use the constant stream of information to adopt the role of the product's lifelong innovator, service designer, intellectual minder and perpetual tutor – following the novel business and operational models that digitized economies enforce.

Why is all this happening? The backdrop is rampant technological progress – quite simply, the most dynamic phase of technological invention and innovation in the 100,000 years modern humans have been wandering the earth. In no time we will be entering an era in which the following marvels will be as mundane as watching a kettle boil: autonomous fleets of unmanned taxis; personalized medical pills that communicate back to pharmacologists and doctors; home lighting that senses your emotions and adjusts accordingly; roads laid using 3D printing; and industrial assembly lines reconfigured in minutes and on their own. Digital high technology, involved in virtually every corner of economic and societal life, will become the norm and play a role in everything we do.

The smart connected product is something entirely new. Its intelligent life is breathed into it via a stack of cutting-edge technologies that have already economically attractive levels of maturity and affordability, or are going to soon. The product of the future is smart because it has cognitive technology on board: software and intelligent algorithms that form a 'mind' allowing for independent decision making. It is smart because sufficient processing power and storage capacity can now be crammed into the tiniest hardware space in almost any device. It is smart because the necessary electronic power supply, batteries or photovoltaic cells have breached the sound barrier in terms of miniaturization, longevity and performance. And it is smart because information on its usage and performance can seamlessly flow between its makers, its users or third parties. This last capability is possible because the product is connected around the clock, wherever it is. It lives on a powerful leash of mobile bandwidth that enables real-time data exchange with cloud and edge servers. There, sophisticated data analytics software creates insights from the relayed data that help the product to show its intelligence while in use and its makers to improve it incessantly.

Finally, the smart product knows what is going on because it has optical, haptical and audio sensors all over it, providing sharp perceptual capability comparable to what humans are born with. The combination of these capabilities will also enable products of the future to be highly customizable and personalizeable in a way that is simply not possible today.

And the technology shows no sign of standing still. On the horizon are new stacks and components about to reach maturity, all of which will also play a role in the smart connected product – among them, higher-speed 5G networks, quantum computing, 4D printing, nanotechnologies, smart materials and bio-molecules.

Special mention must be given to artificial intelligence (AI). In the technology stack for smart products, it is *the* breakthrough capability that is about to push hardware into another league, accelerating the tilt to smart products across the board. While the various technologies that actually make up AI, such as machine learning or natural language processing, have existed for years, they are currently going through such a steep phase of maturation that AI will soon form the intellectual bedrock of all smart products, from home speakers to autopilots in cars to collaborative robots.

All this has deep consequences for how devices are made and put to use. The technology involved will change production cost structures, development styles, and innovation approaches within the remit of product makers as well as their broader spectrum of supply chain and ecosystem partners.

But that's not all. It will also change user expectations for products. Smart products are containers for fluidly reconfigurable software and digital intelligence. As such, they can be adaptable, reconfigurable, responsive and easy to use via exceedingly user-friendly interfaces anytime. All product users from families and individual consumers to industrial and white-collar workers and business leaders will become accustomed to highly tailored 'evergreen' products that can precisely fulfil very individualized purposes at any given moment. There will no longer be large mass markets for uniformly featured products.

Product X.0: a product becomes a service with experience

With all the adaptability and responsiveness, users start to expect more and more complex outcomes from products. A smart mining truck can have a full set of top-notch safety features on board, but the real point will be to orchestrate them to meet precise outcomes agreed between product maker and buyer – say to reduce accident rates by 40 per cent. This will be the real value to the customer and will be what makes the difference in the marketplaces of the future, the so-called 'outcome economy'. The ability to deliver outcomes – lower costs, increased revenues, improved environmental impact – will be the unique selling point of smart connected products going forward. Outcomes, along with great product experiences will be the most valuable currency of the digital age.

In fact, in many cases it will be the outcome that is sold, with the product just the delivery device. Many smart connected products will experience a dramatic shift to as-a-service and outcome-based business models. These models will have the deepest potential for value creation both for manufacturers and users. Eventually, many product makers will not be able to stay in a market without reinvented or newly developed products that enable and support an as-a-service business model. Manufacturers who shirk this will see old passive devices becoming commoditized and turning into low-margin basket cases over time.

Innovation in the new world of smart connected products will largely take place via software updates and real-time, intelligent usage of the data generated by the product. Managers of smart connected product manufacture will need to build in agility throughout the product lifespan because, whether the product is installed in the field or still in development on some drawing board, it is effectively always in development, always updatable by its manufacturer via software. This will have a huge effect on the work of research and development, and on marketing and customer services units. Product profiles and service bundles sold on the back of software updates need careful framing by marketers because individual client groups will become much smaller and more diverse.

Platforms and ecosystems: the new product habitats

For the vast majority of product makers, to deliver the finished outcome and maximize value will, in most cases, require them to work with new technology, services and multiple other ecosystem partners. Few individual companies will be able to provide on their own all the components, software and services needed to run on or in tandem with the hardware product. So, product makers must learn to build entire new ecosystems around their products or partner with and plug into a leading ecosystem created by another player.

Closely intertwined with such ecosystems is the concept of platforms. Typically, the way a platform works is that external partners congregate around a product to run their own business on its back. An agricultural machinery producer might think to data-enable and refine its tractors through making weather data part of their operation. A software company might consider linking up hardware fleets of all sorts to create smart ecosystems that deliver outcomes such as transport or accommodation. Platforms are a source of tremendous value-creation potential, and hardware product makers can become platforms players, although not all can succeed nor should they try.

Reinvention mandate: engineering in the New

To successfully create the next generation of smart connected products and services, companies will need to build entirely new capabilities and get used to running multiple business models and new product roadmaps in the future.

Here in a nutshell are the recommendations this book will lay out in more detail over the coming chapters – recommendations for any maker of hardware or software who embarks on the digitization journey.

Apply digital technology at scale and at pace to make your internal functions as seamless as possible. In many areas within your business organization, teams and individuals will be able to make productive use of the rich and insightful data sets reported back by your smart connected products. They must all be able to get hold of that data and work with it. Only digitization will create the speed and agility needed to supervise, reconfigure and update intelligent products successfully after shipping so that they keep their adaptive qualities for the user across their lifespan. The wholesale digitization of business organizations is what will create the efficiency needed to accumulate the funds to launch the pivot to the new product world.

Always think through your customer proposition from the end point: the user. He or she will show more and more demand for bundled outcomes rather than just for the means, tools and devices to create the desired outcome on their own. To ideate, create and deliver convenient and hyper-contextualized end-to-end experiences is an art in which not only electrical, mechanical or software engineers should have a say. It is also a core task for designers, manufacturing engineers, service technicians, IT professionals, and marketers.

Also, be aware that your product might create much more value for you once it is used in a platform or ecosystem context. Selling smart connected products as mere hardware devices is no longer the only option, and as-a-service models and integration into outcome solutions may well net more value. Thinking through business models should therefore always entail the evaluation of all potential ecosystem arrangements, while practical planning must foresee the right interfaces to interact with ecosystem or platform partners.

Conceptualizing, curating and continually updating smart products after shipping cannot be achieved through traditional skills and professional profiles. A lot of new expertise will be needed. For example, new skill sets such as those of experience designers and platform developers are necessary to master the creation of attractive services and user interfaces. Also, managers must defer decision making to the smaller agile and flexible teams working closely on the products rather than trying to run development processes from the top down.

Committed investment and the allocation of resources between the old core business and the new need perpetual recalibration. The phasing in of the new product world in a business will happen gradually, though this process must constantly be backed by clear entrepreneurial conviction.

This is the essence of the age of 'Product X.0' that we can see rising fast, if we put on the lens that this book is offering.

How to use this book

The book has four parts. Part One sets the scene for the journey towards the smart connected product. It shows how data-driven products will dramatically change and build on customer expectations and even the workings of whole economies and market models, and it explains how businesses can make the most of these trends.

In Part Two, an introductory chapter describes 10 defining traits of smart connected products and contrasts them with traditional products. A new analytical framework – we call it the 'Product Reinvention Grid' – is suggested to target new value spaces, which can scale and mitigate the slowing growth of the core. This grid shows in what way product makers can combine different stages of technological advancement with various stages of user experience sophistication. A product's Intelligence Quotient (IQ) is squared with its Experience Quotient (EQ), resulting in its 'Product Reinvention Quotient' (PRQ). Progressing within this binary grid transforms products into exponential change with them becoming intelligent and eventually autonomous. After that we identify five big shifts pushing businesses in the journey from passive traditional products to smart connected products. This highlights the evolution from features to experiences, from hardware functionality to product as a service, from product to platform, from mechatronics to artificial intelligence, and from linear to agile 'Engineering in the New'.

Part Three focuses on the most important capabilities product-making businesses need to develop and introduces a clear roadmap for their journey to smart connected products. It also provides four in-depth case studies portraying real businesses in transition towards the New and a selection of expert interviews on the topic with business practitioners and academics, revealing further thought-leading insights in this exciting and challenging journey to new product horizons.

Finally, Part Four caps the book by propelling the reader into 2030 where a day-to-day world is highlighted that is for consumers and industrial users almost entirely driven by smart connected products.

Ultimately, *Reinventing the Product* makes a stringent case for companies' need to rethink their product strategy and their product road map along digital lines. Digital technology is simultaneously friend and foe, highly disruptive, and cannot be ignored. Companies that fail to make use of it put themselves in the line of fire for disintermediation or even eradication. But digital technology is also the biggest opportunity in a long time to reposition incumbent product-making businesses internally and externally via product ranges, reimagined to draw from massive new pools of value potential.

PART ONE

Enter the New: smart connected products for the digital age

1

The digital transformation of product making – happening faster than you think!

CHAPTER SUMMARY

This chapter describes the broad transformation process going on in all product-making companies as a result of digitization. It describes how businesses must think, act and eventually become digital along their whole value chain, from the product ideation process to users operating smart connected products in the field. The need is identified for businesses to include all their functions in efficient data loops so as to align them with increasingly connected and intelligent hardware products that work for users 'as a service'. Six digital imperatives must be executed to create growth and value in a highly digitized world.

The digital disruption and transformation of the business sphere is one of the world's megatrends, affecting business-to-business companies representing two-thirds of global gross domestic product. Product makers from sectors such as automotive, industrial equipment, A&D, medical devices, high tech and consumer goods are all undergoing waves of technological upheaval. Their digital reconfiguration, which we call 'Industry X.0', has profound ramifications for their cost structures, relationships with customers, work process designs, innovation drives, human workforce processes and, crucially, for the very essence of their products and services.

This broad-based digital revolution in business is not just about driving operational efficiencies in enterprises. This is what many businesses are doing already but is only the first step towards a digital end game much further down the line. Real digital transformation casts the net much wider. It entails completely new digital business set-ups and ways of working across all company functions, as well as the creation of holistic new operating models around reinvented intelligent products.

These moves are already palpable in many quarters of the product-making sector. Embedded in the wider trend towards the Internet of Things (IoT), industrial and other businesses have started to digitally orchestrate factory floors, workers, enterprise functions, and processes, with some initial success. The number of multipurpose industrial robots has dramatically increased in recent years. In the future these robots will become intelligent, which means able to adapt, communicate and interact. This will enable further productivity leaps for companies, having a profound effect on cost structures, the skills landscape and production sites.

The efficiency gains and net financial extra value created are huge. Overall, manufacturers leading in the usage of digital technologies have so far reported a 20 to 30 per cent increase in gross margins and a 15 to 20 per cent growth in operating income.[1] According to some estimates, industrial businesses can expect a gross increase of their return on capital employed (ROCE) of 25 percentage points by 2035 following the transition to smart manufacturing processes.[2]

Perhaps the most visible driver of this radical transformation is the smartened-up and data-driven product. In the past, the vast majority of products sold were not connected and were fairly passive. These traditional products were usually sold through a third-party channel and the product maker had only a very limited ongoing relationship with the end user.

All this is about to change. Companies are going to need to create smart connected products, what we also call 'living products'.

Four traits distinguish these future products. One, they are connected to the cloud and often directly to other devices. Two, they are prepped for smartness with on-board processing capability and various sensors. Three, they are able to learn using artificial intelligence, voice recognition and other cognitive technologies. And four, many will no longer be sold as a product at all, but rather via an outcome-based, 'as-a-service' business model.

Successful smart connected products will contain densely knit software and digital technology 'tissues' that connect the physical product to the user, the cloud and often a wider community. This software, coupled with increasing intelligence, will be routine in the product-making sector and spawn a distinctive new economic phase. Enterprises will depart from a long-held focus on manufacturing, for broad and anonymous markets, static and passive hardware objects that are at serious risk of becoming low-margin commodities. Instead, businesses will form personalized service relationships with customers, driven by the latter's demand for connected, software-enabled, responsive and adaptive devices delivering 'living' outcomes in real time. This will be true in the ordinary consumer market but also within industry, where such outcomes will become vital to deliver new levels of efficiency and innovation.

The estimated value created by this seminal shift to smart connected products is impressive. According to calculations we made in cooperation with the World Economic Forum, companies and society will unlock US $100 trillion from digital transformation over the coming decades. The digitalization of consumer industries alone could unlock more than $10 trillion for industry and society in just the next 10 years.[3]

A striking individual example of this shift's value creation potential is the startling success of Apple. When Steve Jobs returned to the company in

1997, it was worth less than $3 billion and on the brink of bankruptcy. It saved itself – and then some – by leading the shift to a new generation of connected and increasingly intelligent digital devices and services. In August 2018, it became the first company on the planet to achieve a $1 trillion market valuation – a value increase of more than 33,000 per cent.[4]

Digital eclipses hardware as a value source

Underpinning the requirement to fundamentally reinvent products are the dramatic shifts in the sources of value creation. These shifts began over the last few decades as software started to take an increasing share of product value, but we expect this value migration shift to accelerate in a digital world. Consider the following figures. The sources of value in a typical product currently are: software 40 per cent, electronics 30 per cent, mechanical parts 20 per cent, and digital components 10 per cent. Digital components include artificial intelligence capabilities like machine learning, voice assistant user interfaces, and natural language processing as well as analytics capabilities to capture and process large amounts of data.

In the future, we estimate the value breakdown will be: software 20 per cent, electronics 5 per cent, mechanical parts 5 per cent, and digital

Figure 1.1 Embedded software and digital technologies become the sources of value

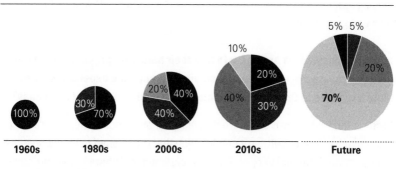

| 1960s | 1980s | 2000s | 2010s | Future |

Source of Value: ■ Mechanical ■ Electronic ■ Embedded Software ▪ Digital Technology

70 per cent. This represents a dramatic shift in value which requires a fundamental reinvention of every product.[5]

An interesting sectoral example: in the automotive industry, roughly 90 per cent of a vehicle's value was once locked up in hardware components – power trains, the suspension, the body, the interior – while roughly 10 per cent was in software and control modules. As we go forward, we will see that ratio shift dramatically to a balance where 50 per cent of the vehicle's value will be covered by hardware and the other half will be divided between software and digital experiences.[6]

The advent of the smartphone changed things in a similar way in the phone market, where value is migrating to the apps and online services rather than the devices themselves.

SMART VOICES   A product executive of an internet product and services giant

How would you describe the role of hardware in a smart and connected product strategy?

It is a role that is not identical in all cases. Microsoft, Amazon and Google see hardware – be it made by them or by third parties – more as delivery devices for services rather than as the centre of their profit generation as is the case with Apple. Services are their main source of value, in the form of the delivery of goods, selling software licences or selling ads as in the case of Facebook. The hardware has here a mere functional delivery role rather than a real value-creation role. In those business models it is often sold at a loss to spread it as widely as possible.

Industrial businesses are trying to turn their hardware products into mere containers for software-delivered services or outcome experiences. Are there examples in the industrial sector you would already call a success?

In the industrial sphere so far Tesla comes to my mind. It tries to create an attractive hardware–software combination that is designed to have consumer appeal through the delivery of a superb experience. Others in the car sector try to mimic that, some with external partners from Google or from Microsoft, but without, I feel, any convincing success yet in the market.

The shift for a traditional industrial manufacturer to set up a smart connected product strategy is more challenging than for a software business. Would you agree?

The journey towards a connected product is no doubt difficult to undertake for a traditional hardware manufacturer. You are in this situation where you have the choice. Either you are disrupted or you throw yourself into this new market. If you decide on the latter you will not get it on the cheap. You will not get it by saying, 'Let's create that little digital unit there'. This will not scale. It will fail; you will fire the team and eventually say, 'It is not for me'. And then you will be disrupted.

Following this trend, all sorts of products will in the end be fundamentally re-architected. Hardware will remain important, but in many cases, the hardware will become a simple shell with functionalities while software and digital technologies become the connective tissue and 'life' elements for value creation.

A defining feature of this product digitization megatrend will be the blurring of the business-to-business (B2B) and business-to-consumer (B2C) sectors. Business-facing enterprises will suddenly have to think in consumer-facing ways as a form of 'industrial consumerism' sets in among business customers. The top criteria of quality for clients will be the final outcome experience and service pervasiveness that helps them drive efficiency in their own operations and growth throughout their businesses. This expectation alone – far more than pure technological hardware features as selling points – will confer success or failure in future digital product markets.

Another defining trend will be companies having to immerse themselves in ecosystems and alliances with partners who today might seem unlikely. Creating smart and connected products on their own will not work; the digital world is too fluid to be handled alone.

This is a radical pivot, a drastic departure from old-style, hardware-focused product making. Managed with care, skill and creativity, it will deliver enormous value. Business leaders will need to think laterally, factoring in a wide and often unconventional variety of possible allies and opportunities. Companies will also need to act with increased speed and agility as the pace of innovation accelerates.

Dual needs: digital transformation and product reinvention

Product-making enterprises must pursue digital transformation on two fronts simultaneously.

On the one hand, they must leverage new digital technologies to bolster *internal* efficiency throughout business functions. The main aim here is to gain as much flexibility as possible in reacting to the fast-moving markets unleashed by smart connected products. In these markets, identifiable customer groups calling for increasingly tailored products or services will become ever smaller and the data insights they give away more focused. Businesses working with that data must build an internal structure able to react quickly and fluidly to such granularity.

On the other hand, digitization must be applied *externally* through the invention of new connected products with smart software lifeblood. This will create new markets, enable new business models and drive market value creation.

As should be becoming clear, then, these dual drives of internal and external digitization must work hand in hand. Most product companies still run very profitable core product or service lines and these should not be discarded but renovated through transformational steps to enable the internal innovation drive. In a separate strand, a second innovation engine for thinking outside the box must be kick-started to come up with new and visionary data-based customer propositions unrelated to the old world. This is where the real reinvention of the traditional product as a smart connected product must occur.

For both battlegrounds – inside and outside the business organization – the adoption of new technologies is key. The dizzying digital vortex is an ever-more-complex mix of underlying technologies such as sensors, cloud computing, processing power, business intelligence algorithms, robots, artificial intelligence, cognitive computing, and big data. Lest there be any doubt, it is entirely this digital technology explosion that has created the opportunity for unconventional growth. Industrial businesses that successfully embrace it to execute their digital strategy around smart connected products will reap unheard-of speeds of profitable expansion.

The product and service development units must become more agile and enabled to react close to real time to changes in the market. And, as we have indicated, in a demand-driven economy, they must be able to quickly hyper-personalize a product or service down to even a lot of just one.

Digitization can spectacularly reduce time to market for manufacturers, as shown by the businesses already investing heavily in this new world. Tyre maker Michelin has cut its time to market down to three years from seven.[7] Even more impressively, the French electrical and industrial equipment maker Schneider has shortened its product innovation cycle from three years to just eight months and is aiming to reduce it still further.[8] Chinese domestic appliances group Haier is now rushing out high-quality products in just 30 days.[9]

All the agility necessary for this stupendous acceleration of processes and thinking is technology-born and data insight-based. The businesses in question achieved it by combining customer data, their own enterprise's data, and the data produced by smart products in the field. We mentioned earlier the increasing granularity that would result from such knowledge. In some cases, this will eventually mean the formerly anonymous mass market shrinks from products of many to personal products of one. This requires a new, demand-driven approach to innovation, one open to external input from extended ecosystems and banking heavily on digital feedback loops with the end-user market. Innovation agility has become the top agenda point, as confirmed to us by the Head of Innovation at Samsung's home appliances division: 'We need to be so quick as time has shrunk and clocks have sped up. What once was a day of 24 hours has gone down to about three hours today. Human awareness, adoption, consumption and disposal of everything has become so fast that one day is worth three days in 1980s calibration.'[10]

The necessary agility and acceleration are only achievable when 'siloed' units within the enterprise are broken up. Unhindered information loops must connect designers with engineers with data scientists with marketers with suppliers with boardrooms with customers in the field. The well-digitized businesses of the future will be defined by maximum data pervasiveness and much more decentralized decision-making processes based on localized data analysis. Data inputs from customers, subcontractors, partners and suppliers must also be channelled to continually influence strategy throughout a product's lifecycle and to arrive at satisfactory time to capability, agility to assemble and speed to market. Again, we see the importance of the ecosystem.

Navigating the disruption: six digital imperatives

Disruption is happening in all sectors, made unavoidable by massive technological change, increasing industrial consumerization and societal pull. Established power plays will be disrupted and industry lines blurred. There are six digital imperatives industrial businesses must harness to successfully navigate this change:

1 **Transform the core.** Digitize and integrate engineering, production and support functions to both arrive at new efficiencies and speed the pace of innovation. This will not only help fund growth in the core as well as the journey into the New; it will also be an operational prerequisite for managing smart connected products and bringing agility, flexibility and acceleration into your whole product value chain.

2 **Focus on experiences and outcomes.** Shift the focus from product features as the basis of differentiation to the end-to-end experience. Create hyper-personalized value to differentiate and lead in the market. This is what markets, consumers and enterprises alike will demand.

3 Build or join an ecosystem. Build, reconfigure, or join ecosystems of relevant partners for the creation and management of smart connected products. No company can provide all the necessary inputs on their own.

4 Work with new business models. Smart connected products do not need to be sold as products at all. Many leaders will shift to as-a-service and outcome-based business models, and all companies should invent and carve out for themselves new revenue streams and value sources.

5 Build a digital-ready workforce. Source, enable, upskill, protect and support the next generation of talent. You will need a new breed of managers, developers and designers for the reinvention of the smart connected product. Embrace the application of artificial intelligence and support the transition to a world where humans and machines interact seamlessly.

6 Manage multiple pivots wisely. Continually balance investment and resource allocation between the core and the New. The entry into the world of smart connected products won't be navigated by simply flipping a switch but will happen gradually. You need to have a clear commitment to be perpetually, incrementally rebalancing your business towards the New.

Figure 1.2 Digital transformation – six imperatives

TRANSFORM THE CORE
Digitize and integrate engineering, manufacturing and product support for new efficiencies.

FOCUS ON EXPERIENCES AND OUTCOMES
Create hyper-personalized value to differentiate and lead in the market.

RE-ARCHITECT THE NEW ECOSYSTEM
Assemble and refresh the right partners to drive new innovation and new capabilities.

INNOVATE NEW BUSINESS MODELS
Invent new revenue streams for new sources of value.

BUILD A DIGITAL-READY WORKFORCE
Source, enable, upskill and protect the next generation of talent.

MANAGE THE WISE PIVOTS
Continually balance investment and resource allocation between the core and the New.

Different sectors, different degrees of disruption

In order to give an empirical picture of where various product-making sector industries currently stand, Accenture has created a Disruptability Index. It measures the degree to which each industry is affected by disruption and what its future disruptive potential is.

Not all industries are exposed to disruption equally at a given point in time. But the vast majority, close to 75 per cent, are either at risk, or are already seeing significant disruption. We see four periods of disruption and each one requires a specific strategy to manage and harness the change. We analysed 3,269 companies across 20 industry sectors and 98 segments, and looked at 15 factors to gauge both the current level of disruption and susceptibility to future disruption.[11] Figure 1.4 shows the results in four quadrants or four periods of disrupion.

Figure 1.3   Disruption pattern definition

	High	
Current Level of Disruption	**Viability** Characterized by embryonic or reborn industries that have endured a significant disruption. High rates of innovation mean sources of competitive advantage are often short-lived, and disruption is constant.	**Volatility** Previously strong barriers to entry have eroded; what were sources of strength become weaknesses. Large disruptors exploit these weaknesses to unlock value, and further compress incumbents.
	Durability Efficient, mature industries, in which structural incumbent advantages pose significant barriers to entry. Incumbent's performance is consistent, and relatively few disruptors are attracted.	**Vulnerability** Structural inefficiencies and low levels of innovation lead to low industry productivity. While profitability is compressed, continued barriers preclude significant disruptor penetration; for now.
	Low	

Low ⟵ **Susceptibility to Future Disruption** ⟶ High

Figure 1.4 Not a single industry is immune to digital disruption

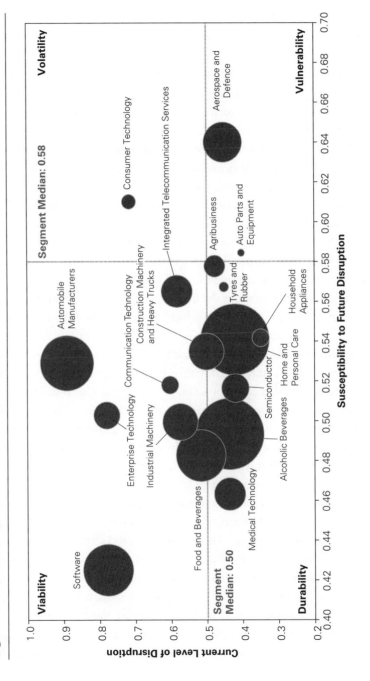

The findings indicate that in major segments of practically all product-making sectors, disruption has set in and incumbent businesses are being forced to react with wide-ranging business model restructures. At the same time, the chart also shows that some sectors such as industrial machinery, construction equipment, chemicals and even consumer goods still enjoy relative comfort. They all occupy the bottom-left quadrant, meaning their degree of disruption and susceptibility to disruption is relatively low so far. Contrastingly, automotive, enterprise technology and telecommunications are already suffering high levels of disruption and offer massive potential for more.

Software occupies an interesting position within this analytical framework, so far suffering the most intense disruption of all, to such a degree that there is not much further disruptive potential left. No wonder software businesses are trailblazers of new digital business models such as platforms and service bundles.

Let's consider in more detail a selection of manufacturing sectors – the car industry, for instance, which we ranked the 12th most disrupted industry. Its main vulnerability is in innovation, which is relatively slow. Spending on advanced technologies at the incumbent companies is still relatively low. A number of high-value start-ups are causing headaches for many established companies, as are various tech-driven giants in adjacent sectors focusing on new modes of mobility. The manufacturer that originally pushed the car most in the direction of a smart connected product is Tesla, with vehicle hardware that can be fluidly reconfigured remotely through software updates, but many other traditional car companies are now closing the gap.

We found industrial equipment to be the 13th most disrupted sector. It scores in the middle range or better for innovation, efficiency and defences against disruption. These relatively good scores are mainly thanks to a lack of prominent disruptors, but the sector has made great strides with platform models around smart connected products. Think for instance of Caterpillar, the heavy equipment maker, which connected large parts of its hardware product range involving third parties.

Even aerospace and defence are beginning to experience disruption and will experience significantly more in the future, in the shape of increasingly ubiquitous drone-based or self-guided flying systems. This sector finds itself in the bottom-right corner of the chart. Goldman Sachs analysis sees a cumulative $100 billion 'total addressable market' for drones during 2016–2021.[12] And plane manufacturers are racing to develop artificial intelligence that will one day enable computers to fly planes without human beings at the controls.[13]

So how to react and position a business, depending on the quadrant it sits in? Here is the gist for each quadrant.

If your business finds itself in the Durability field, put emphasis on transforming the core. Try to build and sustain sources of competitive advantage. Take steps to maintain competitive cost structures. Leverage technologies that create efficiency benefits. And channel investment capacity to extensive experiments to increase relevance.

If you see yourself in the Vulnerability quadrant, focus on scaling the New. Address structural productivity challenges and attend to specific challenges within your core business. Then leverage technology and data to build enhanced services that alleviate customer pain points and build a comprehensive innovation architecture.

Some of you will find yourselves in Volatility territory. If so, pivot wisely. Change course decisively, repositioning the business, and strike the delicate balance between core and New. Avoid obsolescence by pivoting – but not too quickly as this will mean financially stretching yourself too thin. Pivoting wisely is of the essence.

Finally, some of you will fall into the Viability category. Then grow the core, remain in a constant state of reinvention, and direct investment capacity to build new capabilities. Open up incremental sources of growth for the core business by activating new demand and invest in aggressive expansion into adjacent markets.

This sector-wise analysis shows that industry segments find themselves at different stages of disruption. All industries will be disrupted by the pervasiveness of digital technologies and not one industry will be immune to this transformation. Hence business as usual is not an option and although business leaders have largely taken that on board, their reactions are largely piecemeal. They plug gaps here and there but lack an overall masterplan that gets smart products and internal functions to work together over time in a highly connected and software-driven product world.

There is a clear need then for more comprehensive and anticipatory entrepreneurial thinking around digital. Both non-digital business as usual and half-baked digital strategies will likely be fatal for many product-making incumbents. But there are still huge opportunities to get this right. As long as your business is still in the bottom-left corner of the chart, your journey is just beginning and you have every opportunity to successfully master the journey to the New.

Takeaways

1 No industry is immune to the pervasiveness of digital. More than 75 per cent of industries are either at risk or are already significantly disrupted.

2 Digital is also rapidly overtaking hardware as the source of value in products. Companies need to follow the dual-drive innovation approach to digitally transform the core business while simultaneously creating a new breed of smart connected products.

3 There are six digital imperatives to navigate the change: transform the core; focus on experiences and outcomes; build or join ecosystems; innovate new business models; build a digital-ready workforce; and manage the wise pivots across your business.

2
Trends driving the case for product reinvention

CHAPTER SUMMARY

....

This chapter introduces several concepts in more detail that are defining the new digitized era borne in on the back of the smart connected product. It explains the importance of the outcome economy and describes the trend towards hyper-personalization and the supremacy of experience.

....

Trends driving the case for product reinvention

CHAPTER SUMMARY

The arrival of the outcome economy is perhaps the most drastic paradigm shift observable in modern industrial history. It is symbiotically linked to the rise of the smart connected product: only software-intelligent products delivering smart services allow an advanced-stage outcome economy to work and grow, but once you have such products, the growth of the outcome economy becomes inevitable.

What is an outcome economy? In essence, it is an economy in which businesses trade not in the products that deliver certain results, but in the results themselves – the outcomes, such as experiences, efficiencies or safety among many others. Figure 2.1, jointly conceptualized with the World Economic Forum, shows how we get there and where the trend is leading.

From a microeconomic perspective, the outcome economy is a matured phase of the industrial society. In it, consuming households and trading businesses, driven by unprecedented waves of technological innovation, are starting to widen their long-held notions and expectations of what a physical product should provide.

Figure 2.1   **Making the transition to services and outcomes**

Operational Efficiency	New Products, Services and Business Models	Outcome-Based Economy	Autonomous Pull Economy
Digital technologies			
• Improved Asset Utilization & Tracking • Operational Cost Reduction • Equipment Productivity Enhancement • Improved Worker Productivity, Safety and Working Conditions • Remote Equipment Operation	• New 'as-a-Service' Business Models (eg Pay-per-Use) • Product/Service Hybrids • Software-Based Services • Data Monetization • Open APIs/ Developer Networks	• Pay-per-Outcome Models • Shared Risk • New Connected Ecosystems • Platform-Enabled Data Marketplaces • Products-as-a- Platform • Industry Boundaries Blur	• Continuous Demand-Sensing • Perceptive Response • Integrated Human-Machine/ Robot Workforce • Digital Labour • End-to-End Automation • Resource Optimization and Waste Reduction
From Products to	**Services**	**Outcomes**	**Autonomy**

SOURCE © Accenture based on World Economic Forum (2015) Industrial Internet of Things: Unleashing the Potential of Connected Products and Services, January [online] http://www3. weforum.org/docs/WEFUSA_IndustrialInternet_Report2015.pdf [accessed 11 October 2018]

As technology increases the capacity for product connectivity and responsiveness, actors on both sides of consumer and industrial markets begin to think of products in terms not of ownership but use. Hardware products come to be seen as a mere means of delivering services founded on the product's in-built digital intelligence. Shifting from a product-centric to an as-a-service business model is a massively challenging transformation for most companies.

From output to outcome

The output-based economy has been the prevalent economic model for capitalist societies since modern manufacturing and the division of labour came about 200 years ago. Its focus is on quantity of products and services produced. By contrast, in an outcome economy value is defined by the benefits delivered to the consumer or user of the products and services.

A striking, commonly used allegory is as follows: anyone wanting to put up a picture needs a drill to make a hole in the wall, a plug, a screw, a screwdriver, a hook, and perhaps some wire. The different components can either be provided in the traditional way, bit by bit, or the task can be solved through wholesale delivery of the final outcome, a hung picture. In other words, either one rents or buys a drill and screwdriver, gets hold of the necessary ironmongery and plug and does the job oneself, or one commissions a specialized, service-bundling provider and buys the outcome of a picture on the wall at an all-inclusive price.

In mature market economies, this form of market interaction, in which labour, goods and service markets are bundled to become outcomes, is taking hold in more and more of today's seller–buyer relationships, especially affecting the industrial sector.

Perspectives are really changing. In the old-style output economy, industrial business leaders think in categories such as discrete sales figures, conventional unit costs and unit margins achievable with a certain amount of discretionary produced output items – say 500,000 sedan cars sold per year or 1,000 kettles manufactured per day.

In the new-style outcome economy, CEOs of the same industrial businesses will still be diligent profit-seeking managers, but in a different way.

They will face outcome demands such as 'move 3,000 commuters per hour in cars', or 'prepare hot water for tea in 250 households daily', or 'print out and send 1 million customer invoices at a fixed cost per invoice'. They will need a far deeper understanding of their customers' needs. Digital technology and data is key for them to achieve that and deliver these outcomes.

There are a growing number of outcomes in the industrial sphere and we will look at many of them in this book. Most of our examples derive from traditional products reinvented as smart connected products. For now, take the case of Michelin, a business that has manufactured tyres for the last 130 years. The company will soon be offering the following outcome to its clients: 1.5 litres less diesel per truck, per 100 kilometres.[1]

The promised service, labelled Effifuel, has a complex delivery mechanism. Michelin is fitting trucks from commercial logistic fleets with on-board telematics that track drivers' gear-change patterns. The data is used to govern optimal driving behaviour in relation to road conditions, weather and geography. A further part of the outcome package is tailored driver training. Clearly, this is far from the traditional business focus of a tyre maker, and is heavily based on technology.

Perfectly illustrating what we have said so far about outcomes, Effifuel aims to improve economic performance for Michelin's direct clients. This is achieved by the delivery of a comprehensive hardware–software package that helps the buyer to control cost. What's more, the supplier, Michelin, embeds itself in its client's value chain, using technology to turn a traditional product into a product-to-service-to-outcome proposition. Michelin thus creates a whole new customer experience. To be able to bring this outcome to market it has had to change its inner workings completely.

A further up-and-running example is engine maker Rolls-Royce, which is pioneering 'power-by-the-hour' in the marine field and claims the model could reduce customers' maintenance costs by as much as 25 per cent over a 10- to 15-year contract.[2]

The outcome focus is shifting traditional hardware business models to new as-a-service pricing models. Increasingly complex outcome services will very often be paid for by subscription. In other cases, the number of events, time of usage or efficiency delivered might become the basis for invoicing. Charging for outcomes will be much trickier than in the old-style output economy. Business risk will shift to the manufacturers. This will partly be because of the dramatic increase in the number of parameters outside the outcome provider. Ecosystem partners will need to deliver their pieces of the outcome reliably and high user expectations will lead to a very rigorous assessment of outcome quality.

Consider Kaeser Kompressoren, a German business that provides air-compression systems to manufacturing, chemical processing and other industrial companies. Recently it has made the transition from selling cylinders of compressed air to selling, on a subscription basis, what its customers really want: reliable compressed air.[3]

In the printing industry, companies like HP have started offering printing 'as a service' since 2009. Company and consumer clients pay a price per page that includes all the costs of hardware, ink and services.[4]

The software industry is arguably the most advanced industry in making the rapid migration towards a more outcome-based, as-a-service model. Several new entrants were born as 'software-as-a-service' companies, including Salesforce.com and NetSuite. Microsoft has transitioned its products Outlook and Office to a cloud-based as-a-service offering paid for monthly by businesses and consumers, and most other traditional software companies are also in the midst of this transition, including Adobe, SAP and Dassault Systèmes.

It is important to emphasize that, for all the risks, this shift to outcome-focused and 'as a service' has the potential to produce more value for manufacturers of hardware products than an old-style output economy ever could. Instead of fighting for market share for one-off sales, industrial producers will be able to guarantee long-term relationships, and, with those, lasting new revenue streams.

Value shifts across the chain

The rise of the outcome economy is changing roles in the value chain, for example, changing the distribution of infrastructure ownership, freeing customers from the burden of ownership costs and upgrades, as the outcomes or smart services they depend on become based on ever-more-complex technological set-ups. Capital expenditure needs will shift massively from the customer side of the market to the providers of outcomes and smart services.

In some digital business models, a software platform provider acts as an intermediary who shifts the capital expense for the end customer to another provider. The ride-hailing app Uber, for example, has shifted the capital expense of a car purchase from the end customer to the professional drivers.

Analysts estimate that autonomous, software-intelligent cars will reduce vehicle ownership by half. Shared driverless cars fulfilling outcomes to get people and goods from A to B will replace scores of traditionally owned vehicles. According to some estimates, one autonomous shared car will subsume seven existing vehicles.[5]

Market research firm, ReThinkX, even more dramatically predicts a whopping 70 per cent reduction in consumer demand for new vehicles as a result of shared ownership.[6] Furthermore, autonomous traffic could ultimately free up more than 250 million hours of commuting time a year, unlocking a new so-called 'passenger economy'. The chip maker

Figure 2.2 Shift of value in automotive

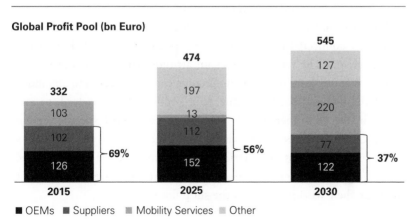

Global Profit Pool (bn Euro)

■ OEMs ■ Suppliers ■ Mobility Services ■ Other

Intel released a study in June 2018 estimating that such an economy could generate as much as $800 billion by 2035, and as much as $7 trillion by 2050.[7]

The automotive sector is already being significantly hit by this ownership shift. Access to usage is quickly trumping car ownership in consumers' eyes, with new upstart ride-hailing firms, built around powerful user platforms, such as Uber, Visa, Lyft, Moovel or Glide quickly gaining momentum. We reckon that, although the overall value pool in the car or mobility sector will double in the next 10 to 15 years, traditional car manufacturers' market share will drop from today's 69 per cent to 36 per cent in 2030 – less than two car cycles away. No wonder big car producers such as Daimler, BMW or Renault Nissan have already poured huge investment into their own mobility platform businesses.[8]

Industrial consumerism, simplicity and ease at any time

The simplicity for the customer in the outcome economy naturally whets the appetite for more. Buying and enjoying an outcome becomes such an uncomplicated thing to do that expectations will quickly rise with regards to delivery standards and reliability.

Business analysts therefore see the rise of the outcome economy as intimately connected to the advent of a phenomenon called 'industrial consumerism'. The term describes an attitude where clients eventually assume that any kind of bundled outcome can be delivered in any industry at the highest speed, on time, to a good quality standard, to anywhere in the consumer world. These concerns eclipse specific technological feature profiles. Customers will also become used to seamlessness as the standard across boundaries that still exist today. They will expect a fluid, responsive digital environment whether they are in their home, their car, their workplace or on the shop floor.

SMART VOICES   An ABB Manager[9]

ABB is an engineering group catering for business-to-business clients worldwide. How are your customers' views of you changing in the dawning digital era?

A lot has changed here in terms of our clients' perceptions in the wake of more and more involvement with digital technology. What the likes of Apple and Google have brought into the lives of all of us, all the daily smart experiences and support through intelligent apps and functionalities, has triggered a real shift in perspective. What has taken hold in the consumer world is now rapidly spreading in the B2B world as well – the expectation to be a provider of actionable outcomes delivered in a fast and uncomplicated manner.

What are the service elements your customers foremost expect?

Much more than in the past, the expectation comes down to, 'Ok, ABB, we have a problem. There is intelligence out there. We expect a solution from you.' That is what our customers, and their end customers, more and more want from us. So, they have stopped caring about our product portfolio or internal set-ups; they want a solution from us, because they have learned that everything can be provided in fluid configurations in no time – like on a smartphone. That is what the standard customer expectation seems to have become, even in the B2B arena.

The phenomenon you describe is framed by analysts as 'industrial consumerism'. What does it mean for your customer proposition?

For us as a legacy provider of complex engineering solutions – for instance in the energy or raw materials sectors – there is an urgent need to step into the shoes of our customer and really understand what they want to have resolved and then arrange our portfolio, knowledge and resources around it. That is where the real new growth opportunity lies for us. We therefore saw some core traits in our business model that had to be broken if we really want to become digitally transformed.

Think of teenagers who expect to be connected to the Internet wherever they are, as if the world is one single Wi-Fi hotspot. In a similar way, industrial manufacturers' customers will become used to such high-outcome standards that the slightest malfunction or hiccough will turn them drastically against the outcome provider.

However, once businesses prove they can deliver complex outcomes profitably and gain acceptance in the market, it will drive their competitors to do the same. And as the outcome economy gains ground, more and more makers of physical hardware will look at ways to attach or embed similar measurable smart services to their products.

Accelerating the pace of innovation

In such a world, businesses must fundamentally change the pace and way in which they conceptualize products. They will no longer try to spot potential market niches five to seven years in advance via conventional market research. Instead, they will use digitally agile market sensing techniques to come up with a product or smart service design in direct and short-termed response to instantaneous demand, even for small lot numbers.

If a business knows enough about a customer's needs, usage habits and preferences thanks to data-enabled hardware products, highly tailored services and outcomes can be designed and offered exclusively to that customer. But that calls for an organization that is agile along its whole value chain, able to handle a fickle market that is constantly sending signals and data that allow for a service to be optimized.

'The traditional hardware development process is broken', Steve Myers, the CEO of Mindtribe tells us. The core goal of new innovation processes, he goes on, is to get experiential prototypes in front of people to allow them to experience the product in the most real-use context as quickly as possible. In that light, design and R&D teams need to operate with maximum agility in such an environment, as software-enabled and upgradeable products will no longer have a stable state for long. Instead they will change their configuration depending on customer data streams. So, designs, technologically as well as aesthetically, will need to be highly flexible. Where, in the old days, teams were dominated by mechanical engineers, they now need strong input from software engineers, experience designers and user-interface specialists.

The way products are designed, engineered, manufactured and supported when in use will change. The work of research and development teams will be less and less predictable as manufacturers will be forced to embed new and not yet fully mature technologies to keep up with fast-moving consumer markets.

In practical terms, this means formerly linear engineering will become more and more looped and agile, often sprinted forward at different clock speeds when it comes to conceptualizing hardware and software elements or designing user experiences. What is more, engineering will

Figure 2.3 Rotating to service innovation

		TRADITIONAL INNOVATION	SERVICE INNOVATION
		Rotating to the NEW	
What?	?	Understand **Customer Usage** and **Expected Product Attributes**	Design and Live **Customer Experience/Journey**
How?	?	Leverage **Traditional and Robust Processes**	Perform **Iterative Design and Prototyping** (to Test, Fail, Learn and Rebound Quickly)
Who?	!	Leverage **Companies' Distinctive Forces and Expertise** around Product/ Service	Manage an **Open Ecosystem** and Perform **Open Innovation** – Acquiring/Partnering with New Talents
Core Skills	✓	**Traditional Product/ Service Know-how** is 'at the Heart'	**Design Thinking and Big Data/Analytics** are the Heart
Motivation		Perform Innovation Cycle in **Years**	Perform Innovation Cycle in **Weeks/Months**

also need to open to the outside and let innovation flow into the organization and into partner ecosystems to leverage off-the-shelf solutions and ally capabilities as and when required.

The US video streaming platform Netflix uses analytics and code written by mathematicians with programming experience to define clusters of movies and connect them to customer movie rankings, factoring in current website behaviour, to ensure a personalized web page for each visiting customer. It further uses data to develop, license and market new content to build a business model around narrow casting, a personalized experience for each of its subscribers. Now that Netflix is solidly in the business of creating new entertainment, the company has also used analytics to predict whether a TV show will be a hit with audiences before it is produced.[10]

You can apply the same principles to a vehicle cockpit so that it will adapt to your driving needs and behaviour. Bear in mind that around 60 per cent of the features in a car are never used, so the space for personalization of features based on gathered data is huge.[11] Or an industrial machine can adjust to a worker's attention levels for maximum safety.

That means that new functions are created as and when needed. Data now enjoys the rank of a strategic production factor. Next to technology, staff and capital, data will become an indispensable core asset for industrial manufacturing and a strategic tool in detecting the user pulse and delivering a service.

To be prepared for such markets, businesses must arm themselves with information technology experts. Data analytics teams must collect, manage and rehash enormous amounts of often unstructured data to distil findings that can be used for updated services or entirely new outcomes and predictive demand propositions. This will create a completely new business function in most industrial companies. If product makers do not do it, their customers will develop their own solutions and capture all the buried digital value.

Take a moment to note how this approach has already become a reality for a few product-making companies. Chinese domestic appliances giant Haier, for instance, has declared the consumer world its research and development department. Based on this, it has overturned current operating models and disrupted relations between business, employees and product users. In the conventional model, product-using consumers are guided by employees, who, in turn, follow management's orders. In Haier's new set-up, built to a principle called *rendanheyi*, the company is effectively led by customers. Employees look exclusively to customer demands before taking decisions. They are paid strictly for the value they create for the end user rather than for the company.[12]

Haier has taken customer-centric business practice into a new league. In the traditional model of domestic appliance makers, customers closed a one-time deal with a manufacturer, buying a fridge or an oven. In the new model, users interact with the business constantly. In the old world, domestic appliances went through wholesalers and retailers before ending up in a household. Haier has cut out these middlemen and delivers the best possible end-to-end user experience by staying permanently in touch with customers via digital channels, interacting with users online, involving them in the whole process of product design, development, manufacturing and marketing. An extended description of their novel business model will appear later in this book in a detailed case study.

The power of personalized experiences

As well as heralding the demise of the output economy, the outcome economy ushers in a new era of hyper-personalization as well as hyper-contextualization, ending more than a century of industrial mass production that had only limited scope for personalized products and services. In the new product world, the starting points for creating products are still use cases. But they will be much more detailed in the early ideation phases, and the product will then evolve further to user contexts once in use.

The era of hyper-personalization is already well underway in consumer markets. For a long time, portable phones were just that: phones. But as smartphones they have rapidly become personalized lifestyle assistants. This was initially enabled by the phones' platform nature, allowing users to download a wide variety of applications and content from a central store. The devices have since become even more personalized with the introduction of voice assistants, such as Apple's Siri or Google Assistant, which learn an individual's preferences and speech patterns over time.

In a similar way, car manufacturers started mass-producing about 100 years ago and soon went on to differentiate production into distinct model series catering to different customer segments along income brackets, geographies, driving behaviour, and family status. But this market segmentation never went beyond the production of a certain minimum lot to keep production of a specific tailored series profitable.

Too much personalization was ruled out both because of overall cost and production feasibility, and because market research had its own limits. Studying focus groups and doing consumer surveys never yielded more than a fairly general picture of what an assumed group of consumers wanted and how customers could be segmented.

However, if, for example, a connected engine can tell the manufacturer something about a driver's behaviour on the road, software-enabled brake systems can be developed to precisely match the driving style of each individual customer and maintenance intervals can be optimized. Or when on-board sensors remotely detect which family member is at the wheel, the manufacturer can put together an appropriate playlist on the car stereo. A car key could be programmed to allow a certain driver to reach only a preset speed limit. Turn a car – or any product for that matter – into a data-linked hardware item and you can all of a sudden make a personalized 'lot of one' or a 'product of me' a reality.

Assistant functions, infotainment components and engine tuning services could for instance be fully implemented but remain inactive at the beginning of a car's lifecycle and gradually be released by the manufacturer to a subscribing user over the lifespan of the product. This is just one example of how smart connected products can hyper-personalize smart services into attractive experiences.

Multiple digital technologies now enable a product to be smart, connected, adaptive and responsive while in use. Of all of them, in our view, sensors combined with artificial intelligence elements such as voice assistant are the most important technologies for building hyper-personalized experiences.

Just think how many sensors are built into a current smartphone to make it adaptive in its functions. Consumers expect it to locate where it is, to know the way it is held, to point out directions, to detect when it is dark, and to identify the fingerprints and faces of users. All that is done by typically around 14 sensors connected to intelligent software – among them, proximity and light sensors, gyroscopes, accelerometers, magnetometers and barometers, as well as monitors for humidity or heart rates.[13]

But the real power of personalization comes from combining sensors and improved connectivity with powerful artificial intelligence (AI) technologies such as voice assistants. Indeed, voice is a real game changer.

French automotive supplier Faurecia is about to go down that route. It will launch a smart vehicle cockpit and cabin interior of a completely new category. Designed to detect who the passengers are and know what they expect, the concept includes, among other features, Amazon's voice-recognizing assistant Alexa as the main control device for a range of on-board systems including personalized entertainment elements. New seat designs will be able to recognize drivers and passengers individually and provide novel health and recreational features to increase road safety, while traditional dashboards are replaced by touchscreens.[14]

An endless multitude of personalized services and bundled outcomes can be imagined once car functions produce enough data to be analysed remotely and transformed into real time.

Ecosystems as the new force

As we have seen so far, the outcome economy and the delivery of experience-based products and smart services go hand in hand. But digital technology is not just the great enabler for clients. It also offers outcome providers powerful tools for embedding themselves in the ecosystems, which will be a typical feature of advanced outcome economies.

In fact, a large part of business leaders' management skills will be about building or joining the right ecosystems. Businesses have two basic options here: they suggest and create a partner ecosystem defining the way they work, or they join an existing ecosystem that is already well established with widely accepted software or hardware standards. In the first case, an important additional question to decide will be whether an ecosystem is designed to be open – working with third-party data or hardware for example – or whether it is closed, with hardware and data controlled by the ecosystem orchestrator. In both cases, new opportunities for additional income are created. Data and services provided by smart products and platform ecosystems allow the creation of mutually complementary value-adding services and new revenue streams.

The creation of ecosystems will be the only way to keep up sufficient innovation speed in advanced outcome economies. Only the combined drive of ecosystem partners will give smart connected products and the services they carry the shape and ad-hoc configuration they need to fulfil specific services.

Against this backdrop, many industrial products will be designed to become platforms. The trailblazers here are the likes of Apple and Google. Both businesses created an ecosystem-style developer community around their smartphone operating platforms. The external app creators are the ones imbuing otherwise passive smartphones with active value – to the mutual benefit of all: the developer, the platform owner and the customer. Other product companies should evaluate their ability to replicate this success around cars, mining trucks, jet engines or home technology such as lighting, security, heating systems or refrigerators.

One thing is for sure in our view: any industrial product will at some stage need to get smart and connected and become part of one or multiple ecosystems in order to survive and thrive in its markets. Though not all hardware products will eventually be transformed into platforms, certainly sooner or later every industrial product will be integrated into another product that operates as a platform. Business as usual might seem to be a third option, but it is not viable as a third party will build a platform on top of your products and disintermediate you.

Takeaways

1 We are seeing the rapid rise of the outcome economy both in B2C and B2B.

2 In this new world, value creation is shifting from hardware to service and 'as a service'.

3 The age of mass customization is coming to an end; it will be replaced by the age of personal experience, use cases and context-specific services.

4 As a consequence, a complete re-architecturing of the product-making value chain and a transformation of the product development cycle are necessary.

The digital reinvention of the product

PART TWO

The digital reinvention
of the product

3

A radically new kind of product: adaptive | collaborative | proactive | responsible

CHAPTER SUMMARY

....

In this chapter, we introduce two new tools to deal with the reinvention of the product: our Product Reinvention Grid and the Product Reinvention Quotient (PRQ). It outlines the evolution on two key axes of the grid, the (Product) Intelligence Quotient (IQ) and the (Product) Experience Quotient (EQ), and describes changing traits as products evolve on each axis. The combinatorial effect of both changes leads to what we call the Product Reinvention Quotient (PRQ) for any product or product company. Connected to this line of thought, we outline five big shifts enabling new kinds of products that are reflected in the grid and that will be analysed in more detail in the remaining chapters of Part Two of this book.

....

Being smart and connected sounds a little like humanity's default mode. We can all adapt reasonably well to changing environments thanks to the way our senses work with our cognitive, communicative and physical capabilities. Pretty much everyone can tell the difference between an executive board meeting, a dinner party, and a funeral. We judge what is appropriate and react accordingly.

Smart connected products – whether hardware or software – will eventually master this kind of cognitive versatility as perfectly as humans. That is why they are also often branded 'living' products. They stand out from the disconnected non-smart crowd by uniting, in a coordinating 'mind', the capacities for adaptation, collaboration, decision making and responsiveness. Many old-world products are going to need reinvention along these lines to meet the challenges of the New.

Orchestrated digital technologies such as cloud and edge computing, artificial intelligence (AI), robotics and 5G networks determine each smart product's capacity for adaption, collaboration, proactivity and responsiveness. The art is to find the right configuration to fulfil a user's desire for satisfying outcome experiences, giving the product's maker a competitive edge and a lasting new source of value. This transition to outcome-oriented products and services also creates new requirements and challenges for the initial product design where user-centricity has to be observed and individualized hardware designs created.

The Product Reinvention Grid

Developing smart connected products is a journey. Based on our extensive work with hundreds of clients across a wide variety of industries, we have boiled down this journey into two key dimensions that are shown on the Product Reinvention Grid in Figure 3.1. The most important lodestar on this trip is to find the right combination between a product's Intelligence Quotient (IQ) – the level of smartness, connectedness and cognitive independence – and its Experience Quotient (EQ), reflecting the quality of experience it can offer through its technology

Figure 3.1   The Product Reinvention Grid

			TRADITIONAL PRODUCT	CONNECTED PRODUCT	INTELLIGENT PRODUCT	AUTONOMOUS PRODUCT
EXPERIENCE — OUTCOME	PLATFORM	OPEN ECOSYSTEM		Smart Home Platform	AI-Powered Industrial Robotic Platform & Voice-Controlled Assistant Ecosystem Platform	Intelligent Automotive Cockpit Platform
		CLOSED ECOSYSTEM	No Product's Land	Car-Sharing Platform & Connected Agricultural Platform	Voice-Controlled Assistant Platform	Ride Sharing Network of Autonomous Cars
	PRODUCT AS A SERVICE			Printer and Tyre as a Service	Intelligent Agricultural Equipment as a Service	Robots as a Service (RaaS)
OUTPUT	PRODUCT & SERVICE		Basic Passenger Car with Warranty	Connected Cars plus Remote Services	Smartphone plus AI-Powered Cloud Service	Robotic Pet with Upgradeable Services
	PRODUCT		Wrist Watch, Light Bulb, Printer	Connected Lighting	Self-Learning Chips Car with Assisted Driving Features	Autonomous Car/Tractor

IQ - INTELLIGENCE CONTINUUM
EQ - EXPERIENCE CONTINUUM

stack and functionalities. Square these two dimensions correctly and you will spot pools of new value for your business.

All traditional products start in the bottom-left corner. They have limited IQ as they have no or few sensors, no AI capabilities and are not connected. Likewise, this coincides with a low EQ as these products are typically sold in a transactional way. With no ongoing customer relationship beyond the point of sale, they become sold and are largely forgotten by their manufacturers. Leaving this corner means that the products must evolve either in experience levels and/or level of technological enablement and settle in various points of this continuum for specific market demands. For different sectors, markets and customer groups, different combinations of IQ and EQ might apply. That is why the upper-left-hand corner is far from the desired spot to be for all businesses. It is very difficult to reach as it involves running a fully developed platform business, which should not and cannot be the aim for all product-making companies. Nevertheless, this is a spot promising huge value for the right businesses.

But let us first access this framework in a more systematic way by examining what is happening on each axis independently. As a product progresses up the EQ axis, the depth and breadth of the customer experience intensifies. The overall journey from feature to experience focus will be described in Chapter 4. The first step, which in reality most product companies have already taken, is to augment the product with value-added services. These could be basic services like warranty and product support, or more sophisticated ones such as data services based on the insights coming from a connected product. Sliding the axis upwards, a massive step change arises with the move towards as-a-service models, which focus on outcomes versus outputs. This requires moving from selling a transactional product to designing, selling and supporting an end-to-end lifecycle experience, a crucial milestone analysed in detail in Chapter 5. A further marked change that some, but not all product companies will take is to evolve their product into a platform that connects with a wide range of ecosystem partners, a strategic tack we will explore in depth in Chapter 6.

Along the IQ axis we find the evolution of the technology stack – from traditional to intelligent and potentially an autonomous product. Here as well, crucial step changes can be delineated. A first move is in many cases to make a product connected via some basic sensor technology that generates and sends product data. The next step towards an intelligent product, which we define as one with product-embedded AI capabilities, may seem trivial but typically requires a major product architecture shift as well as fundamental changes to the product development process. We have dedicated Chapter 7 to this major shift.

As already stated, not all products can end their journey in the upper-left corner of the diagram, nor should they. On the contrary, many new products will start somewhere in the middle and stay there as long as their markets are thriving. Still others might start conceptualized for an existence as a complicated platform product while some more traditional products might take longer to leave the bottom-right corner as they cater for still very profitable markets.

It is worth going deeper into the concept of living products within the IQ/EQ framework. These products evolve intelligently over time, through human intervention or by autonomous impulse, generating dynamic experiences which target different user contexts. These are products with the highest IQ on the x-axis that are capable of delivering experiences of the highest levels of EQ.

Marrying the progression of products along the intelligence continuum with their increasing experience levels is a journey during which the combined evolution transforms into exponential change, with products becoming intelligent and autonomous. It is intelligence and autonomy that makes these products eventually come to life. At this stage they become responsive and steadily progress to becoming intuitive and learning.

To deliver best experiences, such products, or their key components, must become platforms capable of attracting and assimilating the participation of players from multiple ecosystems while data becomes the key input, and security and trust the biggest asset.

From traditional to reinvented products: 10 traits

Let's consider how traditional products stack up against new smartened-up ones. We have put together a table showing the 10 defining features of both categories. Four of these traits are associated with movement on the Intelligence Quotient axis and six are more focused on the Experience Quotient. It becomes clear what a departure from the old world the ever-higher content of digital technology and software in devices means in terms of a value-rich 'living' product existence. This will help you get a sense of the latter's new business opportunities.

Figure 3.2   Traditional product vs smart connected product

	TRADITIONAL	SMART CONNECTED PRODUCT	CONTINUUM RELEVANCE
1. 'Always on' with superspeed highway access	No or very low speed/bandwidth connection	'Always on' high-speed/ high-bandwidth connection to the cloud and Between devices	IQ
2. Sensorized for awareness	No/few sensors	Multiple to several hundred sensors capturing up to terabytes of data per day	IQ
3. Smarter than smart	Fairly 'dumb'	Increasing artificial intelligence and processing power at the 'edge' Processing power of basic device can exceed mainframe from 20 years ago	IQ
4. Software eats hardware and digital eats software	Value primarily from the hardware, but software gaining value in last two decades	80% of product value from software and digitally enabled services	IQ
5. Evergreen via upgrade	No or very limited upgradeability	Living product that regularly receives software upgrades adding significant functionality	IQ & EQ
6. Digital age user interface (UI)	Physical controls, keyboard entry or basic guided touchpad	Digital voice-based UI widely adopted with Some products also using gestures, eye movements or augmented reality as UI	EQ
7. Hyper-personalized	No or limited user customization	Automated personalization based on actual usage behaviour preferences and, in Some cases, user's current mood and context adaptive experience hyper-contextualized	EQ
8. A platform for multiple parties	Stand-alone product	Platform with open APIs to enable third-party partners; comes with a robust ecosystem to 'feed' the platform	EQ
9. Embedded in ecosystems	None	Dozens (potentially thousands) of ecosystem partners that co-develop products, build applications for them, leverage their data or service the products	EQ
10. Digital thread as an eternal umbilical cord	Limited linkage between engineering, manufacturing and 'as is' installed data	End-to-end data models and systems enabling ability to compare the 'as designed' to the 'as manufactured' to the 'as is' product over the entire lifecycle	EQ

Traits to increase the Intelligence Quotient

'Always on' with superspeed highway access

First there is connectivity. We saw in the previous chapters the tremendous advances made by this vital ingredient for the smart connected product world. Traditionally there has been either no or very limited connectivity between makers and users of a hardware product, let alone between the products themselves. It was the software industry that eventually pioneered the concept of 'always-on' relationships – between product creators and users of a software solution – via permanent contact through a cloud server. With the arrival of 5G mobile bandwidth, there will be a solid infrastructure to connect any physical product permanently, allowing for quick design iterations, remote servicing, personalization, and bilateral communication between devices.

Only high-performance connectivity bandwidth will allow for the mass use of cloud technology, one of the central pillars of smart connected products. In that regard, connectivity is among the most powerful enablers of the new smart connected product world and has become a prerequisite for products today. Just consider what capacity leap the coming upgrade from 4G to 5G mobile networks entails. 5G networks will run at much higher frequencies and use shorter wavelengths, making the antennae much smaller but still extremely powerful. Each one of them can handle 1,000 more devices per metre compared to 4G infrastructure, with up- and download times being up to 20 times faster.[1] All these parameters will have direct impact on latency and user experience.

But there is still more to connectivity. In the future, much data produced by smart connected products will be processed and analysed by on-board computing capability, a concept called edge computing. It will mean that products can, to a large extent, self-organize and be independent from connectivity to their manufacturers, owners and users.

Concurrently, however, connectivity between different smart products will increase, for instance in the world of domestic appliances, where it will be possible to coordinate your various devices' energy consumption by networking them. In a similar way truck and vehicle fleets, autonomous internal logistics vehicles and cobots can be connected for platooning purposes and collaboration on the shop floor.

Sensorized for awareness

Another key trait of the smart connected product is sensor and awareness technology. In the old world, products had almost no sensors. The odd data feeler might have been used – temperature or pressure sensors for a combustion engine or industrial machinery – but there was no range of high-tech, low-cost, miniature sensors available to enable mass data collection.

This has changed dramatically. Today a typical smartphone carries, for example, more than a dozen sensors, allowing for sophisticated user experiences.[2] Many hardware products, industrial and consumer-facing, now also have large numbers of practically maintenance-free sensors which, often supported by lifelong battery life, have become low-cost, mainstream products. Up to a few terabytes of data can now be easily harvested in a very short time and sent into the cloud for use by the user, the product maker or even the product itself.[3] Figure 3.3 shows how cars can now be densely sensorized for data collection.

A new aircraft engine includes now around 5,000 sensors producing 10 GB of data per second, corresponding to around 844 terabytes per day based on typical usage. Overall there are a good 24,000 sensors on a contemporary plane. But only 2 per cent of the data generated by these sensors is fully used, as most data sets are stuck in the different subsystems, which do not communicate. This led plane maker Airbus to build a digital platform which allows all these systems to communicate, and make use of 100 per cent of the data generated.[4]

Figure 3.3   Sensorization of everyday products

Examples of Sensors in a Passenger Car

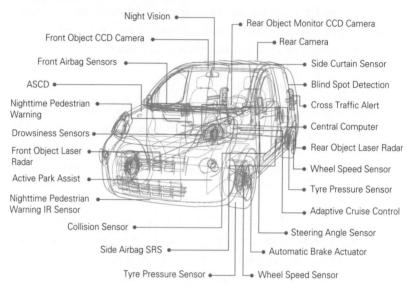

The car cockpit maker Faurecia is developing a sensorized car seat that adapts automatically to personal driver preferences and monitors health data to deliver maximum comfort for the user. Future cars are forecast to carry up to 200 sensors on board – the cockpit alone will have 24, varying from piezo for heat and vibration measurements to weight and position sensors in the seats to cameras – up from 60–100 today, while it is estimated that a typical home will contain around 500 smart devices by 2022.[5] And even in medicine smart sensors are now sent through the intestinal tracts of patients to collect health data for diagnostic purposes.[6] Even washing machines boast around half a dozen sensors these days to control among other things drum speed, water pressure and temperature or balance.[7]

One can see why sensor technology has been such a game changer. However, take note: more sensors mean more complex testing and validation procedures in product development and more operational cost in processing and managing all the data generated.

Smarter than smart

Then there is the new cognitive character of products. Consider how far we've already come: today's smartphones boast at least as much processing capacity as a supercomputer of 20 years ago. From the thoroughly unintelligent, unresponsive lumps of metal, plastic and electronic components they once were, products are emancipating themselves, carrying their own processing, storage and analytics power around with them, turning into 'thinking', autonomously analysing, decentralized, decision-making brains.[8]

Artificial intelligence and edge computing will be crucial technological enablers here, combining with the smart interconnection of objects and cloud computing technologies. More and more intelligence is also being embedded in microchips to cater for distinctive vertical sector applications such as automotive. All this will of course have huge repercussions within the business organizations managing smart connected products.

Soft drink maker Coca-Cola leverages AI to combine weather data, satellite images, information on crop yields, pricing factors, acidity and sweetness ratings to ensure that orange crops are grown in an optimum way, and maintain a consistent taste. The algorithm then finds the best combination of variables in order to match products to local consumer tastes in the 200-plus countries around the world where its products are sold.[9]

Agricultural engineering firm John Deere, meanwhile, acquired Blue River's key technology 'see and spray'. It's a set of cameras fixed onto crop sprayers that uses deep learning to identify plants. If it spots a weed, it'll hit it with pesticide and if it sees a crop, it'll drop some fertilizer. All these parameters can be customized by the farmer, and Blue River claims it can save 'up to 90 per cent' of the volume of chemicals being sprayed, while also reducing labour costs.[10]

In 2012, Amazon acquired Kiva Systems, which develops warehouse robots. AI-controlled Kiva robots have been tasked with product monitoring, replenishment, and order fulfilment. That's a big jump in Amazon's efficiency, compared to the time when humans had to do the grunt work.[11]

Software eats hardware and digital eats software

With the advent of smart connected products, the balance of value will tilt in a transformative way towards software and digital technologies, a shift most dramatically felt by hardware producers. Digital technologies include various types of artificial intelligence such as machine learning, natural language processing and voice assistants as well as the advanced big data and analytics capabilities to harness and utilize all the data being captured by a sensorized, intelligent device. Non-intelligent products are rendered smart and connected; their value profile changes. The pure engineering features lose their market clout to software that makes products adaptable and collaborative.

In the future, electro-mechanical product features will account for only a small fraction of a product's value. The much bigger value driver will be digital data technology built into the product to allow for tailored service offerings and the creation of a product platform to connect to wider fulfilment ecosystems.

That does not mean that hardware is becoming completely unimportant. The key is components – whether hardware or software – that enable a convincing experience for users. Hardware features will still play a part in defining user experiences. They can even be the centre of profit

Figure 3.4   **Simplified view of product evolution**

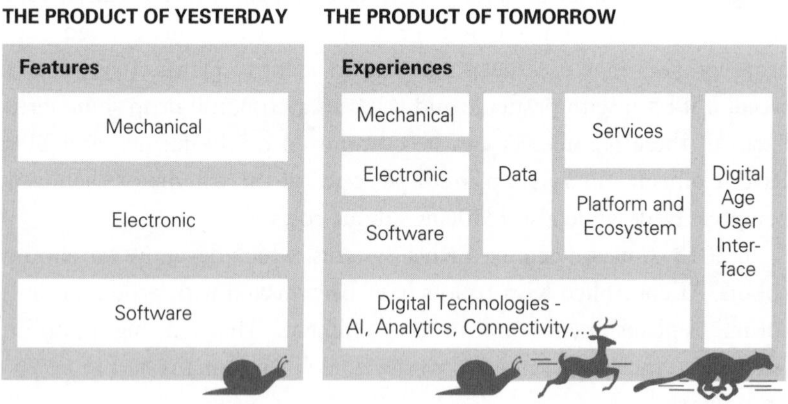

THE PRODUCT OF YESTERDAY	THE PRODUCT OF TOMORROW
Features	**Experiences**
Mechanical	Mechanical / Electronic / Software / Data / Services / Platform and Ecosystem / Digital Age User Interface
Electronic	
Software	Digital Technologies - AI, Analytics, Connectivity,...

generation, as for Apple, or a mere delivery vehicle for services, as for Google for example. Entirely new or heavily re-engineered devices will emerge that depend heavily on how their users perceive their experience profile. Consider for example Haier, the Chinese appliances maker, which has completely reinvented the fridge to make it a social platform in the home. The case is covered in a detailed study later in this book.[12]

Value is not only created by a comfortable user experience, but also through measurable financial impact. An autonomously driving harvester is less expensive to operate than a manned one. An industrial machine that is reconfigurable via software for multidimensional purposes is less capital-intensive than its non-smart counterparts.

To show how seminal the current shift is, let us briefly recapitulate how the reinvention of the product came about.

In the 1960s most hardware products contained practically no advanced electronic components conferring even basic smartness or value beyond the merely mechanical. By the 1980s, however, an average of 30 per cent of a product's value already relied on its electronic components, a share that increased to 40 per cent in the 2000s when software components started to account for around 20 per cent and digital technology elements 10 per cent.[13]

In the future this balance will tilt even more drastically towards digital and software. Technologies such as artificial intelligence or cloud and edge computing as well as connectivity will dominate a product's 'value pie' – with electronic and mechanical features shrinking to a meagre combined value share of just 10 per cent, as we showed in the first chapter. But this might be a challenge bigger than expected for many traditional product makers. As James E Heppelmann, CEO of business software maker PTC told us: 'You don't go to bed as an industrial company and wake up as a software company – it is a lot harder than that.'[14]

It is a big advantage that reinvented smart products, including software products, can add value in multiple areas at the same time. The usage data these products transmit back to base can be directly used to optimize internal processes at the makers, informing for instance research and development units or shaping marketing activities. But exactly the same data can also trigger instant product adaptation while in the user's hands. The same data also helps product makers to remotely service products in the field, again providing insights that feed back into designs for a product's generational updates or create the basis for development of entirely new service propositions.

No wonder it is estimated that smart products will drive global productivity growth of 2.5–5 per cent over the next 10 years. That translates into combined revenue growth and cost savings of $900 billion per year for the industrial sector alone.[15]

Traits to increase the Experience Quotient

Evergreen via upgrade

The capacity to frequently upgrade via software is a characteristic of moving up the EQ axis from a transactional product to an outcome-oriented as-a-service or platform model. Software is fluid: simple code lines can drastically change product characteristics. This gives them an adaptive life, able to offer true experiences rather than mere product features, and to be constantly made new again, literally renewed.

Think of US car maker Tesla. Autonomous driving modes can be enabled overnight via software upgrades.[16] Or think again of your smartphone's operating system, updated regularly to improve usability or data safety. Without this capacity the product cannot react to new user needs or collaborate adequately.

In addition to traditional remote upgrades, newer generations of devices will be able to learn and become real-time aware, allowing them to become both self-configuring and self-repairing.

Digital age user interface

The user interface of a product is the core component of the experience, so to transform the experience most product companies will need to transition to new digital interfaces. In the old days, stationary physical dashboards with a limited number of switches and gauges were the norm. However, product interfaces have now turned into digitally enabled, voice-, swipe- or gesture-activated, artificial intelligence-driven, user-personalized, ergonomically highly adaptable mobile technology modules that allow for seamless communication and collaboration between a user and a smart connected product.

Voice as the core user interface is a seismic technological advance that cannot be overstated. It is the key to a satisfying user experience for any smart connected product. Remember how much this technology has come of age. Where the first on-board voice assistants had to be controlled by defined vocabulary and phrase structures, modern AI-powered voice assistants can fluently deal with natural language commands, making human interaction as easy as talking to another passenger. We very much agree with Nytec CEO Rich Lerz's statement that 'voice is going to have a paramount impact on the future, and is going to unlock and unleash new applications and capabilities beyond our imagination'. Voice, Lerz says, is more unique and powerful than a fingerprint scan, as it can sense emotion and attitudes as well, thus becoming a core element of personalization.[17]

Consider, for example, Nest, which made a thermostat that was one of the first smart connected products to be operated via an in-built touchscreen. Nest's first wave of devices controlled primarily by voice began appearing in 2017. In the near future a wide range of devices will be voice controlled.[18] Gesture controls are being added by BMW and other carmakers as a core interface of the future.[19]

Figure 3.5 **The evolution of human-machine interaction**

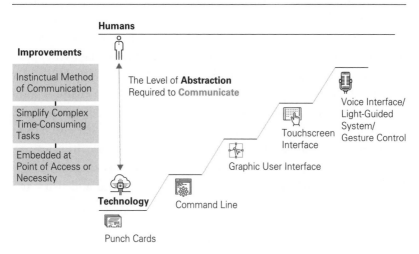

These new interfaces like voice and gestures are much more intuitive and make us as humans a more seamless part of the connected product and service experience. We will communicate with these products and they will in turn communicate back to us in a natural way that does not require learning a new language or method of interaction. Meanwhile, in the background, the product is translating all of these inputs into the digital world inputs.

Hyper-personalized

Once a product has digital user interfaces and enough intelligence, it is possible to consider a degree of personalization that simply was not possible only a few years ago. These new user interfaces are flexible enough to deliver the right user experience (and the right outcomes) at the right time in the right, personalized context. But their software-enabled flexibility is also an important precondition for highly tailored experience personalization in individual user contexts. Having the flexibility and capacity to morph into a very personal product offering for a large number of users is one of the crucial drivers of customer value in the age of smart connected products. Rendered autonomously intelligent by artificial intelligence a product such as a car can even learn and take its own decisions on how best to personalize its user experience.

In car interiors, old-style dashboards are increasingly giving way to uniform touch-sensitive interfaces to allow for maximum personalization. Future car cockpits will be intelligent enough, thanks to sensors and locally acting artificial intelligence, to personalize the user experience to an even greater degree. As described above, the average car will have a wide array of sensors, which will be able to enable a completely new generation of personalized experiences. All cars will be able to identify who is in the driving seat, then adjust seat and mirror positions as well as playlists on the car stereo automatically. Through AI the car also adapts to individual user conditions such as fatigue levels, and to external driving conditions such as traffic density and weather.

A platform for multiple parties

Another decisive development is the impressive transformation from a product in isolation to a connected product platform, allowing for the deep and versatile involvement of third parties. Think of your smartphone, which comes to life for your day-to-day use and experience only by means of the numerous apps provided by third parties. Or consider Faurecia again, with its digital car cockpit primarily operated by users via the voice assistant Alexa, provided by Amazon.[20]

While many smart connected devices will not become platforms as no new value will thereby be tapped, some surely will. Only a platform – be it as a closed proprietary or more open ecosystem-oriented version – opens a product up to really rich adaptability, responsiveness, collaboration and personalization experiences. A platform also opens up the opportunity to extend into new markets and capabilities beyond the traditional core product. For example, Apple is now a leading provider of music, entertainment and other cloud services. In the case of most smart connected products, this will turn them into serious value generators for the manufacturer, the user and the third-party developer.

For this to happen, hardware products will need open APIs to make the co-creation of experiences as easy and financially attractive as possible for partnering external innovators.

Embedded in ecosystems

The emergence of the product as a flexible and living platform goes hand in hand with the emergence of the ecosystem that builds organically around it. Ecosystems mostly emerge because third-party applications are being developed to run the platform, because third parties are approached to service the product, or because they leverage product hardware and/or data to create their own complementary service designs for the product user. Ecosystem developers orbiting a platform today might number anywhere from a few dozen to millions, as in the case of Apple's iOS and AppStore.

The emergence of product ecosystems is one of the defining trends of the new product world in an era of complex digital business models. Yet ecosystems do not involve just third-party developers. They are also the framework for forming alliances with strategic suppliers or external marketers. Product ecosystems are a natural extension of the platform existences of smart connected products and a vital prerequisite for their commercial success. The war between competing products will be won by the better ecosystem bundling the more convincing customer proposition and user experience. Platform product manufacturers must therefore not only enable but also support, nurture and manage this ecosystem.

The emergence of ecosystems around smart connected products poses new challenges for marketers and brand managers. In a hyperconnected product world in which mobile phones, home thermostats, domestic appliances and even sports clothes are increasingly connecting to the Internet and potentially each other, brands have to learn to play well with each other or give up a certain amount of control to those who own the most popular interfaces. In most ecosystems, the brand in contact with the end user will occupy the most lucrative position.

Digital thread as an eternal umbilical cord

In order to provide an outcome-based and compelling experience, it is essential to be able to track and trace the product over its entire lifecycle – including the changes to the hardware, software and data attributes over time. This requires smart connected products to run on a data leash controlled by its makers long after sale, in a way that almost no hardware maker does today. These goals will be achieved via two related concepts, the digital twin and the digital thread.

The digital twin is a complete digital representation of a physical product, including not only 3D modelling, but also the material properties, the software and data. The digital twin becomes the single version of the truth of all product-related master data.

The digital thread extends this concept over the entire product lifecycle to track changes to the product's configuration over time and trace its data flows. With the digital thread, a service technician in the field could compare the 'as is' configuration of today, to the 'as manufactured'

Figure 3.6 The digital twin concept

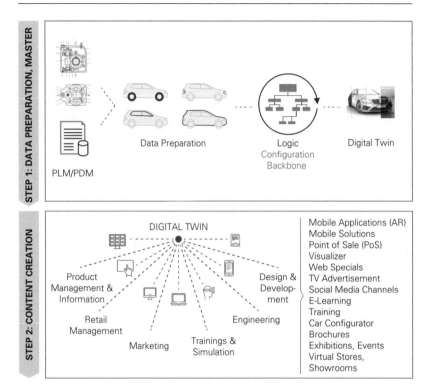

STEP 1: DATA PREPARATION, MASTER

PLM/PDM

Data Preparation

Logic
Configuration
Backbone

Digital Twin

STEP 2: CONTENT CREATION

DIGITAL TWIN

Product
Management &
Information

Retail
Management

Marketing

Trainings &
Simulation

Engineering

Design &
Develop-
ment

Mobile Applications (AR)
Mobile Solutions
Point of Sale (PoS)
Visualizer
Web Specials
TV Advertisement
Social Media Channels
E-Learning
Training
Car Configurator
Brochures
Exhibitions, Events
Virtual Stores,
Showrooms

SOURCE © Mackevision

to the 'as designed' product. If the technician is equipped with augmented reality technology, she can even see these views simultaneously and compare them.

Digital products that are fully instrumented with sensors, data processing capacity and connectivity and that are also backed by a digital twin and digital thread allow multiple streams of data, from feature usage to performance, to be sent back to product development and engineering teams. This kind of 'listening' to a smart product used in the field, and the actions in response to findings and usage patterns it enables, are central to the concept of a smart product that evolves technologically over time via constant new software iterations installed remotely.

The existence of a digital thread platform and subsequent collection and analysis of usage data also opens up new opportunities to monetize the data in new and innovative ways. For example, data collected by a Wi-Fi company on who is accessing the network and where they are physically located could be used to add intelligence to building systems to tell facilities people not to turn off the air conditioning at the normal time of the day because there is a group working. The majority of companies do not have digital twin and digital thread capabilities today, but these will become a critical capability underpinning reinvented products in the near future.

The Product Reinvention Quotient

Returning to our Product Reinvention Grid: by taking the combinatorial impact of moves along both the IQ and EQ axes, we have created an index that we call the 'Product Reinvention Quotient' (PRQ). It is a measure of the degree of change required to achieve the transformation from a traditional product to any quadrant on the matrix shown in Figure 3.7.

We believe there are two main breaks on this matrix that represent discontinuities. The first is the IQ axis move from Connected to Intelligent Product, where we assign a jump from 90 to 120, and the second is the move on the EQ axis from Product & Services to 'Product as a Service', which moves from 30 to 60, or double the complexity. We want to stress that the upper-right-hand quadrant is not the desired outcome for all product companies. Indeed, we believe that very few companies should aspire to this quadrant. We will revisit in detail how to navigate this grid and drive the product reinvention in Chapter 10.

Figure 3.7   The 'Product Reinvention Quotient' = IQ + EQ

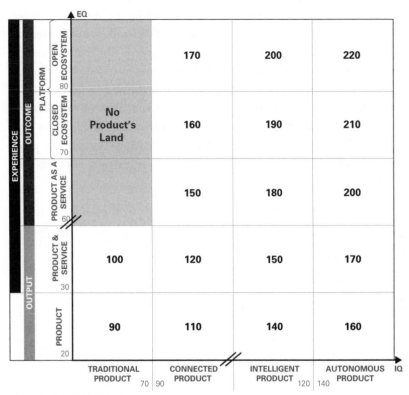

IQ - INTELLIGENCE CONTINUUM
EQ - EXPERIENCE CONTINUUM

The future is here, start now

Virtually every product will need to be reinvented and transformed in the near future. In fact, the race is already on, and business leaders who dither for too long about creating new generations of smart connected products run a significant risk of their organizations being disrupted and even pushed from the market. Indeed, the massive disruption represented by the product reinvention is enabling new entrants, often from completely different sectors. For example, very few consumer electronics companies would have imagined five years ago that Amazon and Google would become two of their most formidable competitors.

But reinventing the product itself can only be the start. Remember, smart connected products, in an advanced stage, will be able to remote control, optimize, adapt, decide and respond entirely by themselves. The resulting business cases will be so radically new as to be almost inconceivable to business leaders today. But all this is just around the corner. In order to be ready for it, it is vital to start your pivot away from the old product world and build by intelligible increments towards the New.

The following chapters provide details on five big shifts that all product companies need to consider in order to successfully reinvent their products. These are:

1 From features to experience.

2 From hardware to 'as a service'.

3 From product to platform.

4 From mechatronics to artificial intelligence (AI).

5 From linear to agile 'engineering in the New'.

Takeaways

1 A new product world is emerging in which products are becoming more intelligent and experience-rich. We can plot every product of this emerging new world against a new analytical tool that we call the Product Reinvention Grid.

2 Companies can manage to drive the increase of the IQ and the EQ. The magnitude of the management effort required to be successful can be estimated by combining the desired shift in IQ and EQ to calculate what we propose as the Product Reinvention Quotient (PRQ).

3 We have identified five big shifts. For most companies, a major transformation exploiting the five big shifts will have to be managed to succeed with their products in the new digitally driven world.

4

Big Shift One: From features to experience

CHAPTER SUMMARY

....

This chapter sets out the transit of businesses from their orientation on product features to the experience-driving product in the environment of an outcome economy. The distinction between mere product features and more fluid, high-value experiences and the difference between B2B and B2C experiences are explained. Finally, empirical data shows what business leaders think on the topic and the economic potential of a consistent experience strategy.

....

As we have already shown to some extent, products change their perceived status for users when they become smart and connected generally and evolve along the EQ axis specifically. To further elaborate: where users expect an inactive product to deliver relatively limited support for specific tasks, an intelligent product carries expectations of reliance, cooperation, adaptability and trust. And from enjoying mere functional features with a passive product, users come to expect a seamless overall experience with a smart connected one.

When a worker picks up a conventional screwdriver, she or he is aware of the need to find a suitable screw, exert appropriate manual force, work to the right entry depth and under adequate light for safety and efficiency. A smart power driver, or even an intelligent screwing cobot 'thinks' all that through independently. It will find solutions for every single issue we thought of, and maybe some we didn't. The smart power driver will also communicate and interact with us in a natural way, for example via a voice UI that does not require reading a lengthy manual or learning a new language. The sum of these answers amounts to a rich, more productive and safer user experience, way beyond the enjoyment of a non-smart screwdriver's bare product features.

Many products have already reached limited smartness and, by this, have bolstered their experience massively. Take the washing powder brand OMO's product Peggy, a smart peg with sensors combining Wi-Fi connectivity and a smartphone app with the ability to monitor temperature, humidity and sunlight. By taking weather data and combining it with local micro weather information, it pushes out reminders to users, telling them when to hang their clothes out and when it's better not to.[1]

We took intellectual input from all our in-house experts in the field, combined with a crowd-sourced survey within our organization to sharpen the concept of user experience. According to the results, good user experience qualities have the following attributes: they must take a deeply personalized, dynamic and emotional shape and work on seamless connections across all touchpoints between user and product. Furthermore, to secure a maximum of convenience, they need to be embeddable into platforms without any noticeable friction, and last but not least, they are best conveyed, enjoyed and supported by adequate technology such as virtual or augmented reality.

The slow but certain waning of the feature economy

We all know from longstanding personal experience how mere product features can be described. A car gets us from A to B with a fuel efficiency of X miles per gallon, a clothes peg holds clothes on a line supporting up to Y kilograms of weight, and a light bulb has Z watts of power output. To think of these examples as more elaborate experiences would work like this: a car gets me from A to B with not more than a click on an app; a peg alerts me when the clothes are dry enough to be put into the wardrobe; a light bulb knows what light colour I prefer when coming home from work.

Over time, the arrival of more and more products with digital intelligence on board will phase out the product feature-led economy and replace it with a more experience-led economy. Consider this in the business-to-business sphere and you see how it not only creates profitable new markets for industrial hardware and software makers, but also delivers massive efficiency gains in the management of industrial assets.

Products that are bolstered by new digital technologies such as artificial intelligence can act like futurologist assistants for their users. They can anticipate a situation or desire and make just the right move on behalf of the user. Smart cars learn about their passengers and provide tailored entertainment experiences. Smart trucks provide pre-emptive gear-change functionalities for a comfortable driving experience. Smart lighting can adjust brightness and colour hue based on the weather outside and the user's mood. A smart lock can open based on recognizing your face.

The productivity gains, the competitive edge, and the potential for improved satisfaction at work should become clear from these use cases. As one example, the aircraft maker Airbus uses smart glasses that point out screw holes for their workers to install plane seats with greater accuracy and speed in its A330 plane model.[2]

Experience: a quantum leap beyond features and services

Characteristically, mere product features and services remain external to an individual user. Traditional product specifications and features are by definition completely non-customized: all customers, users or employees can benefit equally from the same product or service. Contrastingly, experiences, almost by definition, are distinct and personal for each individual and subjectively assessed by them for quality level and ability to adapt. This is why experiences are more difficult to design and carry higher potential for disappointment, but also much higher potential for value creation.

So, what is an experience?

An experience is a pure holistic concept that is based on the sum of all interactions between a consumer or enterprise user and a providing company, its products, services or brands. The perception of an experience is formed by the interaction of the human with their environment and consists of two emotional qualities, arousal and valence, as seen in Figure 4.1.

Figure 4.1 The perception of experience

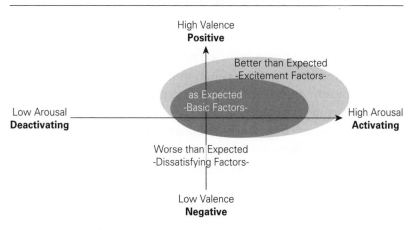

SOURCE © Design Affairs

An experience is formed by processing stimuli coming from the environment. This involves sensory processing, perception, emotion and cognition – conscious and subconscious – and leads to a response, decision or action, which is also part of the experience. The resulting subjective evaluation of an experience – as positive or negative, unimportant or important and so forth – is influenced by environmental factors such as the situational context as well as by factors such as gender, personality and mental states such as attention, fatigue, workload, vigilance, stress. In addition, the motivational state and expectations towards an experience influence its evaluation.

How does this apply to a product?

Referring to product experience, the product is the stimulus, which triggers the information processing and leads to an experience. Again, the emotional quality of the experience is highly dependent on various factors: the context of usage, the task and aim of the usage, as well as the user with their individual personal predispositions. The expectations users form about a future experience with a product are essential. The conscious and emotional response increases exponentially when better or worse than expected factors occur.

In order to form a positive experience, different properties of a product can be influenced – such as ergonomics, functionalities, design, usability – that can be measured and evaluated to define its impact on the user without forgetting to deliver on the basic expected factors. Referring to our EQ categories, experiences can be further analysed. Product experiences can stay closely tied to user experiences with a product, mostly applicable for consumer products. In the case of B2B products the experience can rather be framed as a positive operational outcome, an efficiency or productivity gain, for example. This is again different from an ecosystem experience where the experience is driven by inputs across a whole ecosystem and not exclusively by a defined product.

Figure 4.2   **Framework for the experience formation**

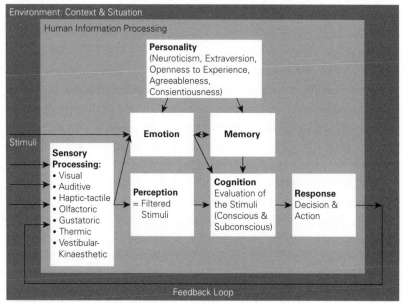

SOURCE © Design Affairs

Experiences are often lumped together with services and even with feature-rich products. But lumping together is just not enough. This is to underestimate the power of the experience. The different components must be designed, architected and seamlessly integrated. Take birthday cakes. With the maturing of the food processing industry, providers increased their offerings from a few standard cake types to offer a rich array of toppings, flavours, imagery and shapes – more features. Then, as the service economy took hold, several cake companies provided cake-as-a-service, with online interfaces allowing you to order an entirely unique cake, even specifying your own recipe. Now a few businesses have gone further and truly expanded into delivering a more integrated experience, becoming event-organizing cake makers to whom an entire birthday celebration can be outsourced. These service companies used cake making to form a relationship much more value laden for both provider and receiver: an experience around the cake product, the organization of a social event centred around food.

Let's look at a similar example in a B2B context. Industrial robots have been around for a long time. As technology evolved, they started becoming feature-rich, then more agile and collaborative, doing different tasks and, most importantly, becoming more flexible and adaptable for different industrial needs. Automakers prefer the flexibility that comes with cobots, which don't need to be in a fixed factory location caged in or cordoned off and can therefore be rolled around to different parts of the factory. You can have them do a task in the morning and an entirely different task at a different location in the afternoon. All well and good, but with robots like that, we're essentially still in the realm of features.

Now, however, on the back of the explosive growth in cloud computing, storage, machine learning and AI, robots as a service (RaaS) have started gaining traction worldwide. Here, customers lease robotic equipment and run it flexibly via the cloud. A blessing to various sectors including manufacturing and agriculture, it makes scaling robot capacity up and down easy and allows buyers to try out robotics without going through a capital-binding purchase process.[3] For the makers, this creates the potential to expand the market dramatically.

In a further step, with artificial intelligence becoming the mainstay of value creation, the industrial world is seeing the emergence of 'robots-as-experience'. Companies like SoftBank Robotics, with their Pepper robot, have demonstrated the technical feasibility of building robots for customer-facing use cases. Pepper is the first of its kind, a humanoid robot capable of recognizing the principal human emotions and adapting its behaviour accordingly.[4] With this kind of offering, robots have come a long way from just being customized hardware–software tools.

Consider also Sony's Aibo robotic dog, which takes the concept of a personalized experience to yet another level. Each Aibo will 'evolve' differently based on how the user/owner 'raises' it. On arrival, Aibo is essentially untrained and, just like a real dog, needs schooling based on the user's voice commands and positive reinforcement. Aibo can even learn to recognize individual faces.[5] More and more B2C products are using this kind of technology and it is only a matter of time before this kind of customization through machine learning and adaptation comes to some B2B products as well.

But also in more traditional equipment-making sectors the concept of rich experiences equating to competitive strength and value is gaining traction. Italian industrial machinery producer Biesse is leveraging AI to develop a new adaptive human–machine interface for their wood, glass and stone processing machines.

The purpose is to ease and enrich the operator's experience based on their evolving behaviours. For this, Biesse developed a software named bSuite, a 3D CAD/CAM suite enabling the company's customers to design solutions to be executed by Biesse's machines. bSuite, trained by the usage of specialized operators, will be able to progressively learn and simplify the user experience, automatically providing an optimized set of patterns. The smart integration of bSuite technology into machines allowed Biesse to gain a competitive advantage in the market of industrial machinery thanks to innovative and wide functionalities based on customers' production and ways of work.[6]

BMW's 'Drive Now' car-sharing scheme is another good example. It offers seamless registration in a minute, full transparency on available cars, and an integrated map to guide you to the car. When you get in the car it is personalized and its navigation is set to your destination.

The increasing importance of experiences is a relatively new phenomenon, driven in large part by the growing penetration of digital devices around the globe. Mass digital connectivity has resolved the problem of reaching potential new markets and consumers have found they use these products almost constantly. Such always-on customers are now demanding really engaging and meaningful content tailored to their needs, as well as ever-more-complex fulfilment experiences – a challenge but also a huge opportunity for all product companies.

The flipside is that in an always-on world any experience faces a harsher spotlight. It is constantly reviewed by its receiver, requiring careful expectation management and permanent, innovative responsiveness from the experience provider. Brands and companies should seek to make their experience as enjoyable and reliable as possible, leveraging the many digital customer touchpoints created through the growing use of digital hardware to keep customers engaged with brands and offerings. That is how they will capture or enhance mind share.

The differences between B2B and B2C experiences

It has long been well known that customer experience is top priority for consumer-facing businesses. Now, business-facing companies have started taking this seriously too. Eighty per cent of B2B executives participating in a recent survey identified customer (user) experience as their key strategic priority.[7] This is clearly because they know mere product features and functions are losing traction and the capacity to create differentiation in the marketplace.

The starting point for the creation of such a compelling experience is to gain a deep understanding of the customer journey, in other words of how customers will find, use and interact with the product over the

Figure 4.3   A B2C example – the customer journey of a camera

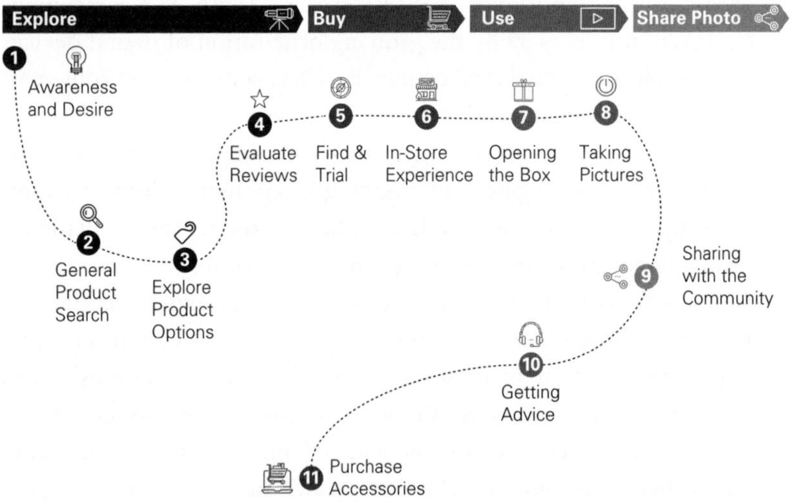

entire lifecycle. Particularly in B2C industries, we have found that often the critical experience innovations are less about the product features and more about simplifying the purchase process and product support. Here is an example of a customer journey taken from the camera industry. It shows how protracted such a journey can be even for a relatively simple consumer product.

Leading companies identify what we refer to as the key 'moments that matter' across the customer journey. They then develop innovative approaches to address these moments. In B2B industries, the journey is often a much more complex one with many different stakeholders to consider including the end user of the product (who happens rarely to be the buyer), the corporate purchase decision maker and the channel partners who distribute and support the product.

In the B2B world the experience happens at enterprise levels. There are multiple touchpoints in parallel, which makes the experience much more complex to manage. Also, the lifecycle of the experience-delivering product or industrial asset is much longer than that of a consumer product. And the industrial product or asset is typically embedded in the customer's business or even in the customer's own products and services, which makes reliability much more critical compared to a consumer product experience. If an 'experience-rich' consumer product suggests a product that is 'distinct and personal' for unique, individual consumers, it seems an 'experience-rich' piece of industrial equipment would mean the equipment is 'distinct and personal' in the way that equipment relates to unique, individual industrial environments – such as new construction sites, new manufactured products or materials, new environmental conditions such as heat, humidity and so forth.

Given the huge risk involved, B2B customers tend to be rational and unemotional when deciding how to invest in things like shop-floor machinery. Good experiences for this group are overwhelmingly utilitarian, with outcome objectives such as improved productivity, operating efficiency, turnaround time or asset utilization. The focus is primarily on the experience's economic benefits, as well as safety and ease of use. It is meant to practically support shop-floor managers, machine operators and inventory staff, and facilitate things like maintenance, repair and upgrades.

Figure 4.4   A B2B partner experience example from the High Tech industry

Pre-Purchase Experience	Purchase Experience	Post-Purchase Experience	Relationship Management
I need to understand what a company does and does not offer. What does it take for me to choose a company? What resources is the company providing me to sell to my customers? I want to be guided on the best way to buy/sell a company's products/services. I need to easily understand which of a company's offerings would meet my/my customer's technical needs, budget, and timeline. I want the company to clearly communicate that they understand my/my customer's needs (spoken & unspoken).	I need to trust that the company has created the complete, relevant, and valid solution that will get the results I want at a competitive price. I expect a company to be predictable and transparent around pricing and availability, including changes. I want a company to be flexible to my needs when creating a payment plan and contract. At the moment I place an order, the company has made a commitment that I expect them to fulfil. I should understand what I am paying for/being paid for and the payment process should be simple.	While I'm waiting for my order, I need the company to reassure me that everything is going to happen as promised. Delivery of my order occurs as committed and the invoice matches what I received. I need the company to guide me and support me through the registration & activation process. If something should go wrong after my order, I expect the company to make it right. After installation, check in with me to confirm the solution is working as anticipated.	Check in with me to confirm the solution is still working. Help me understand what I need to do to keep performance optimal. When something goes wrong, give me a simple path to reach someone who can solve the problem. My/my customer's needs are changing. Help me understand, if the current solution still fits or if I need to upgrade or replace. Keep me updated on the team assigned to meet my needs. Support me to sell more of the company portfolio, leveraging new business models simply and easily.

By strong contrast, a typical business-to-consumer experience (see figure 4.3), say the performance of a smartphone app, a cab service or a fridge replenishment service, includes the emotional, physical and psychological connection a customer has with a company. It is a subjective response to direct and indirect encounters with the company's offerings and representations – from product to services to brand. The experience could span the whole product lifecycle, from brand discovery to research, purchase, use, customer service, maintenance, upgrading, retirement and, finally, product recycling.

It will become even more complex in the future, as consumer experiences at home, in a car and at work become interconnected to deliver a single, seamless experience customized to an individual's lifestyle. It also spans the entire customer lifecycle and includes every touchpoint a customer has with the company. It is therefore much more driven by personalization and convenience.

Yet as we have already seen, both worlds, business-facing and consumer-facing, are tending to blend more and more. If we think, for example, of automotive suppliers such as Faurecia, which we highlight in a case study in this book, we see that their route into the world of smart connected products leads squarely to both segments. Faurecia designs its new smart car interior with an end consumer in mind, but the product itself is sold to carmakers for inclusion in the vehicles they assemble.

Engineering user experience

Experience came out of the fact that stand-alone products and services do not deliver enjoyable and convenient outcomes. The experience economy is where product makers compete not only on how their products work but on how enjoyable they are while working. So, experience became important because companies started competing for the attention of their customers. In an experience economy, once businesses have a touchpoint with a customer, they must think about how to wrap it in an experience to make the customer more engaged. And this is what has created a new and special discipline – experience engineering – among other more traditional engineering fields. It is focused on designing, creating and managing products while they are in use – with the single goal to create an experience for maximum user satisfaction. Experience engineering has implications for the core architecture of many products, as we will discuss in Chapter 5.

Digital experiences for both B2B and B2C are created very differently from feature-rich products and services. Three critical functions illustrate this: design, development and manufacturing. In most industrial practices designing a conventional feature-rich product is limited to hardware features created by electro-mechanical engineers, who control R&D almost entirely. The product design in these circumstances is frozen once complete, until the following product generation starts a new design cycle. Manufacturing – again mainly hardware-oriented – then focuses on the resource-efficient, fault-free assembly of physical components to ensure all features are up to desired standards.

Even when services are ideated and conceptualized around a conventional product, these activities mostly stay focused on hardware, with only the occasional involvement of marketing experts. The primacy of hardware often even hinders practical and financially lucrative product serviceability.

In stark contrast, the design, development and production of wider experiences requires obsessive attention to the holistic interactions across hardware features, software-enabled functionality and increasingly cloud-based services. Interdisciplinary engineering capabilities must be formed in R&D units and more consumer-facing functions such as sales and marketing must be closely involved. Hardware design also becomes modular, quite often leading to a reduction in physical complexity. Instead, software design must enhance interoperability and flexibility so that the product can be updated constantly throughout its lifecycle to continue to provide a relevant and desirable user experience. Essentially, the manufacturing process becomes fluid, continuing via these updates throughout the product lifecycle.

Today's reality seems to stand against this. The head of digital services at an eminent premium carmaker told us that it still needs eight years to take a new car model from ideation into the market, with the final design frozen three years prior to launch. Bound to such slow speeds, it is still one of the largest challenges in the automotive sector to anticipate what kind of service or experience would be perfectly pitched to the market years down the line.

Challenger car firms such as NIO work the other way around. They start out with a user experience, and try as fast as they can to build a car model around it. Direct research on what such an experience could look like was essential, a manager of an automotive start-up in China told us:

'We just asked the user time and again. For our concept we tried to minimize classical market research. Instead we went straight for customer research and focused, for instance, on ethnographic criteria. We selected very small groups of people from our assumed target groups and spoke to them in their own homes and on their own terms.'[8]

Once the customer journey and the 'moments that matter' are understood at a deep level, opportunities for true innovation in the experience will become more readily apparent. But product makers must bear in mind that they need to engineer the whole user experience for all occasions, not just on the ones when the product is in use. It means working on each of the touchpoints or 'moments that matter' mentioned above.

Good and bad experiences

Profitable growth in today's always-on world depends on providing experiences customers will pay for and that retain workforce talent. Satisfied customers and workers alike transform themselves into ambassadors for the experience provider and today their ambassadorships can be digital, harnessing the power of social media to spread the word far and wide. The question arises of what the best-rated experiences are.

While experiences are personal to individuals and entities, and therefore not strictly comparable, one can still differentiate between good and bad on a string of broader criteria. Simply put, three factors are key to a good experience: ease, convenience and speed. This is supported by B2C customers participating in a recent survey conducted in 33 countries and across 11 industries, where 71 per cent see *ease* and *convenience* as two key characteristics of a good customer service, and 61 per cent identify *speed* as a third key attribute. The survey confirms, unsurprisingly, that bad experiences drive customers away, which can cost dearly. Fifty-two per cent of B2C customers in the United States switched providers within the year preceding the survey, due to poor service, at an estimated cost of US $1.6 trillion.[9]

The majority of survey respondents said a good B2B experience is when the business buyer is provided with 'timely insights and support to improve the reliability, performance and safety of its industrial plants, infrastructure and other asset-intensive operations and services'.[10]

The requirement for predictive maintenance is a great case in point. Schneider Electric, the French electric and industrial equipment group, has combined digital and Internet of Things technology with B2C thinking around customer experience, creating its intelligent demand sensing solution.[11]

Analysing huge volumes of customer data on production, consumption and electronic batch processing through a proprietary algorithm, the company can predict equipment failures and take appropriate action well in advance. Based on this intelligence, Schneider's equipment can adapt to its environment at speed, reducing overall downtime and improving asset utilization for clients.

Top talent is likewise vital for delivering top-class experiences. And, as we have shown, to retain the best personnel, organizations must deliver an engaging work experience. What are the key attributes of such an experience? It must inspire innovation and creativity within the workforce. It must forge greater collaboration and help employees realize their organizational and personal goals.

This is easier said than done for large bricks-and-mortar companies heavy on assets and non-millennial staff, but steel group Tata has, we think, found a way forward, fostering tandem relationships between millennial workers – employees aged under 30 – and the experienced leadership team through a reverse mentoring programme. Younger employees spend part of their time bringing senior leaders up to speed on the latest digital trends and technologies. It's a cross-pollination, with the younger employees also learning from their elders' experience and professionalism.[12]

But recall our previous discussion of ecosystems: in today's digital world no single brand can embody an entire experience. Instead, it will be a blend of offerings from various providers. Companies must be innovative in figuring out the feasible combinations. Ikea's Tradfri smart-home lights are now compatible with Apple's HomeKit smart home standard. This means users will be able to control their Ikea smart home lights from the Home app built into Apple's iOS mobile operating system rather than be forced to use the dedicated Ikea app.[13] In the industrial sphere the Siemens MindSphere and SAP Leonardo software packages orchestrate machine communication within smart shop-floor environments. They can be fitted to all kinds of machinery made by various vendors and used in different sectors' manufacturing plants.

Human + machine experiences for the industrial worker

Often overlooked in our view is that workers are also consumers. Like any consumer, they dictate grocery lists to Alexa and query Siri for restaurant suggestions in their private life outside the workplace. So, it's no surprise to see the growing global digital workforce increasingly demanding the integration of intelligent technologies to improve their experiences and productivity at work, a phenomenon some are calling the consumerization of business technologies.

What does experience mean in this context? Here, our previous distinctions between B2B and B2C break down somewhat. Like consumer experience, workforce experience also includes the emotional, physical and the psychological bonding the individual enjoys. A workforce experience is about the subjective response of an individual to an enterprise that directly or indirectly employs him or her. It could begin with job discovery and go on to include the selection process, the whole set of on-the-job experiences, holidays and leisure, retirement, retrenchment and rehire. It also includes 'micro moments' like booking work travel, or mingling with colleagues over lunch, and covers physical health and wellbeing and more abstract measures of health like a sense of belonging and self-actualization. Note that, because each worker judges her or his workplace experience subjectively, the same need for personalization appears here as in B2C experiences. The opportunity arises for businesses to gain a much more granular knowledge of their workers and strengthen workers' personal relationship with the business.

A 2018 global study, based on a survey of 14,000 workers spanning four generations and representing all skill levels across 12 industries, found 68 per cent of highly skilled workers and nearly half of their lower-skilled peers speaking positively about the impact intelligent technologies will have on their work.[14] Until now, robots, big data analytics and other technologies have worked in parallel with people but in automated isolation, improving process efficiencies. Now, as companies invest in intelligent technologies such as AI that can sense, communicate, interpret and learn, businesses can start moving beyond automation to elevate the capabilities of their workers and unlock trapped value from shop floors to store fronts.

Our research shows that if companies invest in Human + Machine collaboration experiences, driven by AI at the same rate as top-performing businesses, they could boost revenues by 38 per cent between 2018 and 2022 – or as much as 50 per cent in the consumer goods and health sectors – and lift global profits by a total of US $4.8 trillion by 2022. For the average S&P 500 company, this equates to US $7.5 billion of revenues and a US $880 million increase in profitability. Our conclusion, naturally, is that making the workforce a firm part of an experience strategy will pay off. Investors agree: companies providing great employee experiences outperformed the S&P by 122 per cent.[15]

What's at stake?

Several studies startlingly reveal the stakes for companies in driving experiences for customers and the workforce. The data shows that B2C high performers with regard to experience can meet or exceed customer expectations 91 per cent of the time – 21 percentage points higher than their peers. High performers achieve higher rates of success across many business criteria, notably return on investment (+14 per cent) and cost savings (+6 per cent), but also differentiation (+22 per cent), improved relevance (+21 per cent), customer satisfaction (+16 per cent), customer loyalty (+17 per cent), revenues (+11 per cent), and scale and efficiency (+22 per cent).[16]

The opportunity cost of under-delivering B2B experiences is substantial. A leading technology manufacturer boosted sales by 4.9 per cent by orchestrating customer retention, contract renewal, product refresh, and cross- and up-sell experiences, while reducing unproductive administrative time for channel sales teams. The research also reveals that companies with effective partner lead generation and coaching are 63 per cent more likely to exceed their indirect channel revenue goals.[17]

Without the right customer experience, B2B companies can't build relationships with loyal customers who will spend more, stay longer and be more forgiving once they can enjoy a good experience. They miss out on growth opportunities that reside outside the traditional sales cycle. And they incur huge opportunity costs by failing to capture renewals.

This is what we took away from numerous interviews with industry practitioners conducted to research this book. They can be found in full in Chapter 11. They gave us broad confirmation that product experiences are now one of the top three strategies business leaders should have in mind in product-making sectors. A top executive of Dassault Systèmes told us:

'Unless you engineer the product's complex user experience from the
beginning, it is going to become extremely difficult downstream to
reconnect it to be truly connected. That is the essential logic behind these
technologies. From the first day you think about a smart connected product,
you don't want to separate development and engineering from product
usage and ultimately the end-to-end experience it provides.'[18]

A manager at Tesla said that one of the most important tasks for the carmaker's designers is 'The development of the user interfaces integrating the vehicle's capabilities to a unique experience for the user',[19] while a executive from a High Tech company confided that it is 'To create an attractive hardware–software combination that is designed to have consumer appeal through the delivery of a superb overall experience'.[20]

Takeaways

1   The basis of differentiation for a smart connected product is not traditional features and functions anymore, but holistic user experience.

2   Designing a compelling experience cannot be an afterthought. It must become an integral part of the product's value proposition and therefore be designed, engineered, monitored and refreshed as such.

3   As any product moves up the Experience Quotient (EQ) continuum, the experience becomes richer, broader and requires a robust ecosystem to power it.

5

Big Shift Two: From hardware to 'as a service'

CHAPTER SUMMARY

This chapter details the implications for product-making companies managing the shift to as-a-service business models as a step towards the outcome-based economy. This is a journey fraught with challenges and requires changes to every business process and function, but offers major rewards to those who succeed. We show that the role models for such companies are software companies and a few leading product companies who have successfully made the switch.

Big Shift Two: From hardware to 'as a service'

'Everything as a service' is becoming the mantra of the future outcome economy.

Combinations of devices and services will be the central business model for product makers. As all product devices become more intelligent and connected they will produce large amounts of data and this will form the basis of a whole range of enriched usage scenarios from which service-based business models can be derived. This will depend, however, on supporting digital infrastructure being developed – within the product in use and at its makers – that can be reconfigured in real time to spawn new value chains and income.

For example, connectivity and smart devices enabled French tyre maker Michelin to move some of its business from selling tyres to ensuring mobility and safety as a service. Via intelligent sensors embedded in the tyres, performance and lifecycle are monitored and coordinated to ensure replacement when needed. This business model was born from a clear recognition that customers value experience and their own business success higher than the mere product features of a tyre set. After all, hardware excellence is a must in today's markets.

The problem is, businesses active in legacy products such as engineering and manufacturing, often decades old, tend to be extraordinarily strong believers in hardware excellence as the basis for their financial wellbeing. In most cases, the market niches they have carved out, occupied and defended against competitors seem to them to guarantee their continuing success. While many have invested in embedded software over the years, they often give this valuable layer away for free today as something hard-bundled with the product. To arrive at the novel mindset in which a hardware product's main value driver is its software and digital capabilities sold 'as a service' versus a transaction product is no easy mental journey for them – let alone adoption of the requisite operational changes.

Software industry leads the way

The software industry is a pioneer in moving from a traditional product-centric business model to an as-a-service model, referred to as SaaS

(Software as a Service). This business model and operational revolution was initially led by pure SaaS start-ups such as Salesforce.com and NetSuite, successfully followed by traditional software players such as Adobe and Microsoft. It's clear where the value is. We analysed industry performance and over a three-year period, traditional software companies had an annual revenue growth of only 5 per cent while pure as-a-service players enjoyed 27 per cent growth.

The traditional software companies that have successfully made the transition to as-a-service providers are particularly interesting to study as role models for industrial companies wanting to go in a similar direction. Adobe, for instance, had an as-a-service sales share of 19 per cent in 2011. After a drastic strategic shift, this had risen to 83 per cent by 2016. Today the company has 8 million subscribers, adding around a million more service customers per quarter. The market has richly rewarded Adobe for making this strategic pivot. Over the last five years, the company's value has grown from $12 billion to around $122 billion (August 2018) and its stock's price to earnings ratio shot from 12 to 58 in the same period.[1]

But many other traditional, product-based software companies have struggled mightily to make this shift, as their traditional processes, go-to-market channels and organizational culture resisted the change. Companies that made the successful pivot, like Adobe, have undertaken a large-scale transformation of all their core processes.

Product as a service

The urge to become a service company and thereby attract investor attention is now permeating every sector. This includes the industrial world, where more and more companies are considering giving selected core products away for free to monetize a product-based service that will, at least in theory, drive more revenue over the lifetime of the customer. Here's a schematic overview of how different industrial companies are tackling this.

Figure 5.1 Products as-a-service business models

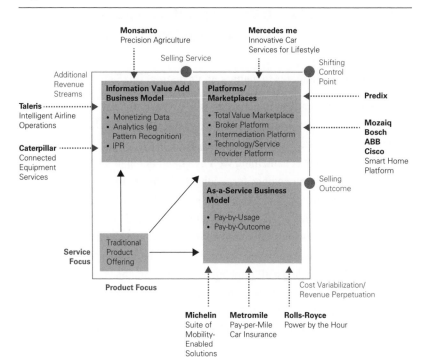

Enterprise hardware makers such as HP and Lenovo have started to run devices-as-a-service business models. Companies that used to sell hardware for medical tests are now looking at managing the data results of the tests and providing analytics-based healthcare based on personal electronic records. The automotive industry, meanwhile, has invested in ride-hailing and car-share operations, offering cars as a service for consumers to use as needed. And several industrial engineering firms have begun to use software to communicate with, diagnose and service electronically controlled engines and machines used in the field, offering maintenance and replacement as a service.

SMART VOICES   Q&A with Bill Avey, Global Head of Personal Systems Services, HP Inc

What gave HP the impetus to embark on its distinct DaaS strategy?

Next year will be the 80th anniversary of HP. There are not many hardware competitors left who have managed to navigate that highly competitive technology space for so long. That can only be done by constant transformation. 'Device as a service' is something we have been doing in the managed print space for business customers for a long time, because we saw this as a decisive competitive edge in a maturing market. That is the main part of the genesis of this model at HP. We eventually extended that model into the consumer print space with schemes like Instant Ink and a subscription model that has been received very well. And we extended it then to our PCs as well. We are currently building a 3D print business and from the very beginning it is being created as a DaaS model to give us the edge in the market. Right now, across all our business, we are finding ways to deliver industry-leading HP innovation as a service. This model plays an important role in our future strategy.

When did HP turn that into a declared company-wide strategy?

About three years ago we were getting more and more demands from our customers saying, 'Hey HP, we really like what you do for us in the managed print arena. We also like your PCs and the way they are designed to go with services. Could you also do in the PC space what you do in managed print services?' They were basically looking for a partner that can provide devices in combination with all sorts of continuous services – beyond the bundle of services and accessories we were offering them anyway at their discreet demand. Based on this new demand, we started what we eventually called 'Device as a Service' in 2016. These comprehensive device management services enable customers to modernize legacy IT environments in a smart, efficient manner. By having HP manage device fleets as a service through-out the entire lifecycle of the technology, IT departments free up valuable time and resources to invest in strategic growth initiatives inside their own organization. It's a win-win.

Expand a little on how that works in practice.

For example, when you get to that moment when one of your PC's batteries needs replacing we would be able to anticipate that before it happened and proactively ship you a battery. Or, say there is an executive assistant who worked in a sales department for which we had initially provided the right device. Then this assistant receives a development opportunity and moves into a marketing department. All of a sudden application software such as Photoshop and Auto CAD becomes part of this person's day with the need for more storage and processing demand on their device. Via our analytics tool we can set the alarm bells to give people what they need at any time. We can say, hey this person had the right device over the past 15 months but three months ago it all changed. The machine is telling us that there is something wrong and we can now say why and go back and actively remediate it.

Lastly, describe the typical customer of your DaaS offering for PCs and what pricing model is typically picked.

I would say that the average services customer is one who is procuring the device from HP, combining that with a standard bunch of device lifecycle services to have it up and running. These services would comprise, next to the managed analytics services I talked about, Windows imaging, managing the right BIOS settings, and physical as well as electronic asset tags. It would also involve the actual deployment of the device, the data transfer from old to new device, and expert support as well as asset recovery of the old machine. All of that would be bundled up into a per-seat-per-month fee. This fee is made up of the services, the device and the term, which for most customers is three years. But these parameters vary. For one customer, a global restaurant chain, we have a per-device-per-month model for the cash registers. Or take our ink subscription programme for home users. It basically allows our consumer home office folks to subscribe on a monthly basis to a certain number of pages printed.

These incumbent companies are waking up to what software pioneers recognized a good while ago: strong customer demand, beyond hardware and traditional product features, for a tailored customer experience. In addition, there is the need to innovate much faster and deploy new capabilities and upgrades frequently, as well as to learn from actual product usage and product telemetry data as direct inputs into the future product roadmap of improvements. Both of these are best accomplished in an as-a-service model.

Redefining the core product architecture

How should products be re-architected for carrying services? Let us first recapitulate that the smart connected product of the future will be responsive, it learns from interactions with its user to deliver customer-centric outcomes, it can become a platform for customers and producers to drive hyper-personalized outcomes, and, importantly, the product is capable of finding the fastest way to drive high-quality user outcomes.

From this vantage point, there are fundamental implications to be observed by manufacturers: AI-driven software is to be embedded to make the hardware intelligent while operating in the field. Software and digitally enabled services have to be installed with the product and a business and payment model defined.

The product architecture is becoming software-driven, which needs to be reflected in the overall product engineering approach. This includes, as Figure 5.2 opposite indicates, software engineering 'inside the box', for instance, by embedding a software stack running the device and another one running the chip. But it also requires software engineering 'outside the box' to enable services, operate applications, or to run platform software that runs independently of the device.

When a company intends to make a quantum leap in performance through its smart connected products in comparison with its old traditional product, it needs to design a completely new product architecture, for which Figure 5.3 (see page 98) gives a comprehensive overview.

Let's highlight briefly the most important components required for the product architectures of the New. The future product is able to

Figure 5.2 The new product engineering stack

Product & Platform Engineering	Innovation & Product Strategy	Device, Services & Experience Roadmap
'Outside the Box'	UI/UX for Software & Platforms	Designing the Experience
	Platform Software	Software that Runs Independent of Device
	Application Software (Product)	Software Built as Part of Product
	Industrial Design	Design the Product
'Inside the Box'	Embedded	Software that Runs the Device
	Silicon/Semi (Front End, Physical Design)	Software that Runs the Chip
'The Box'	Hardware/Mechanical Engineering	Hardware Design

constantly send usage data to help develop the next product or improve the customer experience. The future product also allows for testing and software updates 'over the air'. It requires a secure handling of the data, in particular the confidential user-based data. As it becomes intelligent, the product needs embedded software to operate sensors, semiconductors, actuators. It will be 'always on', hence it requires sufficient connectivity features. The product enables digital services based on AI, big data analytics, machine and deep learning. These intelligence components are driven by external tools and databases, which need access to the product's operating system. The product also builds a customer-specific experience based on adaptable user interfaces.

With huge data loads being generated by sensors and processors in smart devices, a combination of cloud and edge computing will provide the mechanism for storing and managing all this data. But more importantly, it will execute real-time analytics for insights and informed decision making. This will allow product makers to shift to rapid, almost continuous release cycles and iterations as the true power of cloud computing is its ability to transform speed and agility.

Figure 5.3 Smart products architecture

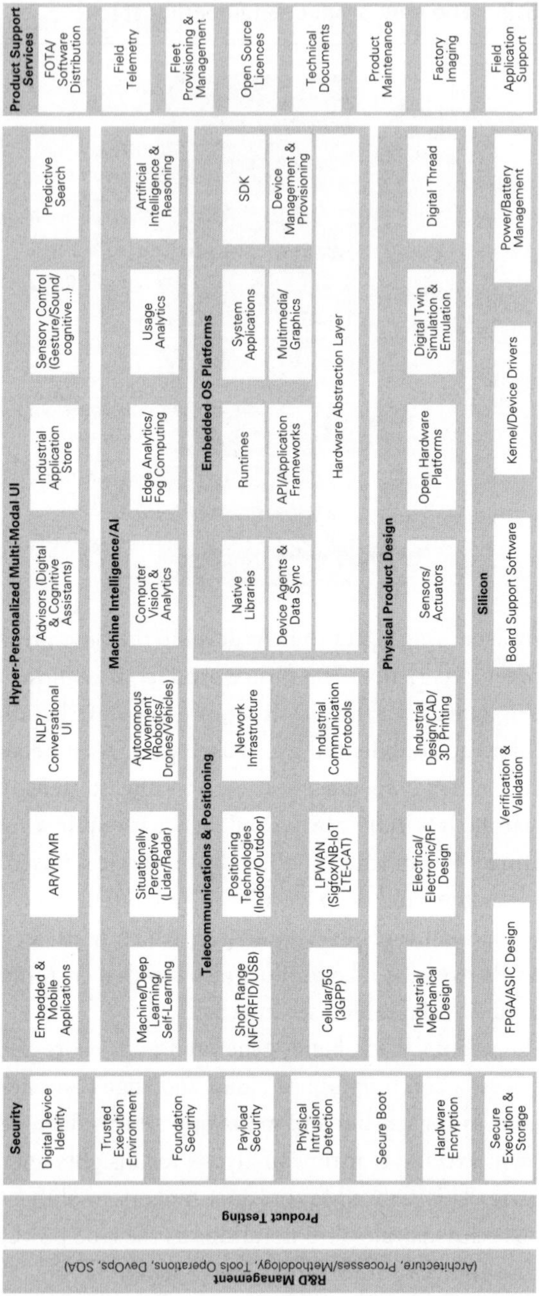

It is an enterprise-wide challenge

The challenge of turning a product-centric company into an as-a-service provider is huge – especially when serving highly demanding business customers. Building the operational capabilities required to serve B2B clients seamlessly as you migrate from the old world to the new one is daunting. You need a proper migration strategy and the means to achieve service-level and other expectations.

In a world where service experience is built around a smart connected product, customer experience becomes doubly important. Customers enjoying a product service expect more and depend on the provider more. Rather than simply buying a passive piece of hardware, service customers expect adaptability and interaction.

And service touchpoints present many more ways for a provider to either satisfy or disappoint. Customers depend on their technology providers as much as they depend on outsourcing providers for key business processes. The consequences of failure for business or brand will be harsh.

So, let's sketch out the key pillars of success here, along with the operational challenges we have observed in companies attempting this transition. All this is as valid for tech and software companies as for industrial businesses.

Figure 5.4   As a service: the five pillars

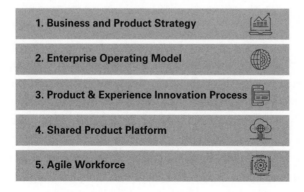

1. Business and Product Strategy

2. Enterprise Operating Model

3. Product & Experience Innovation Process

4. Shared Product Platform

5. Agile Workforce

Figure 5.5 Product to as a service transformation

	Traditional ——————————▶	As a Service
'Product'/ Offer Development	Feature & Function Focus	Customer Experience Based
	Waterfall Development	Continuous, Agile Development
	Engineering 'Silo'	End-to-End Digital Thread and DfX Mindset
Marketing & Sales	Traditional 'Product Marketing'	Digital and Service Marketing Focus
	Channel-focused selling 'Push Model'	End Customer-Centric Selling 'Pull Model'
	One-Time, SKU-Based Pricing	Annuity Pricing Based on Usage or Outcomes
	Sales 'Silo'	Cross-Functional Marketing, Sales and Service Teams
Service & Support	Customer 'Support' Focus	Customer 'Success' Focus – Tracking Outcomes
	No or Limited Usage Tracking	Robust Adoption Tracking & Metrics
	Reactive Break/Fix Support	Proactive Customer Support Core to Offer
Finance	Simple Product Sales Accounting	Usage-Based Invoicing; Services Revenue Recognition
	Low Invoice Volume	High Invoice Volume & Complexity
	Sales & Channel Compensation Upfront	Compensation Based on Customer Lifetime Value
IT	Product-Based IT Systems	Service-Based Systems, Including Subscription Billing
	IT Completely Separate from Engineering	More Integrated IT and Engineering Operations
	Limited Analytics	Robust Analytics & Big Data Core to 'As a Service'

1 Develop a clear business and product strategy – reviewed and supported by the CEO, the senior leadership team and the board. It must include a transition plan since the as-a-service business will typically lower revenues in the short term as the company moves from revenue at point of product sale to the more diffuse revenues of a subscription model or similar. That move goes hand in hand with the transition from capital expenditure to operational expenditure

patterns, which can have positive balance sheet impacts for customers. Where in the old product world businesses were used to spending in individual investment rounds for product projects, in the service world they will keep spending for their permanently run services and balance this with permanently flowing income from service proceeds.

2 Remember that you are not offering a product and its hardware qualities but the fulfilment of an outcome or outcomes in the shape of a service. This requires a transformation of virtually all aspects of your company's operating model, from development to go-to-market to operations. To structure this, bear in mind that, while you will still be delivering a product, it must be perpetually adaptable to customers' service/outcome needs, which should drive everything from upfront ideation and experimentation to systems and process implementation. This is particularly true for both product engineering and operations. And deep engagement with the customer will continue to be required across the entire lifecycle. To drive service usage and adoption, many leading as-a-service companies have created entirely new 'customer success' organizations that work with individual users. This is a completely foreign concept to most traditional product companies, which tend to focus on the upfront product sale, with post-sale support typically centring around repair services when something is broken. Finally, 'customer fitness' teams focus on improving financial performance over time for both the company and the customer.

3 How does product design and development change when creating a product as a service? The short answer is quite like as we explained in Chapter 3; the shift from selling a product to 'as a service' requires a major increase in the EQ score. While for years the best products have been designed with a sharp focus on the feature needs of the user, product companies are suddenly being tasked with designing services, requiring a much deeper understanding of user journeys in order to create a holistic experience. In addition, for services allied with smart connected products, the number of relevant disciplines increases. Product designers and engineers will need to work with experts from

Figure 5.6 Continuous customer engagement

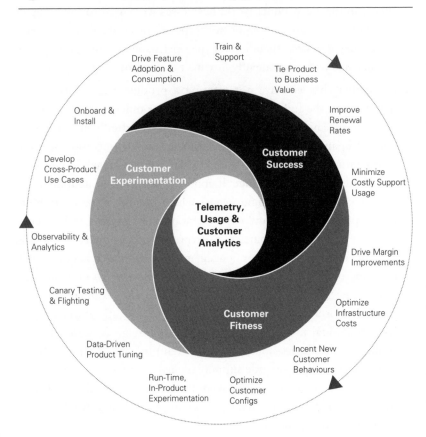

fields such as AI, data analytics, user interface design and more. The other big change is to the innovation cycle. Where in the past a company would have made its one-off user observations and applied them to design a product lasting for years, now it will be designing a digital experience that feeds back, providing the company with new data – new user observations – as soon as it starts being used. This means companies can and in fact must be continuously engaged with their customers; more details about this will be covered in Chapter 8. Finally, in a world of product services it is much easier for a client to switch from one service provider to another. So, designing, supporting and constantly updating services becomes a key business competency with direct impact on the profit and loss accounts.

4 Create a business and product platform to support the as-a-service model. Most product companies have traditionally separated internal information technology and product engineering activities, but in an as-a-service world, these functions must become intertwined. Business functions where products are conceived and eventually made must link up closely with the general data flow throughout the company. For example, once the smart connected product is installed with the client, the data it generates must be connected to the corporate systems to enable monitoring and corrective action. In addition, companies need to develop business platforms that enable customers to quickly and easily add users, upgrade features or adjust service plans. Very few product companies have such capabilities today.

5 The need for improved coordination across all of the various business functions increases dramatically in the as-a-service world. All functions need to adopt agile principles or risk being held back by the slowest function. Remember, your product will be changing permanently while in use in the field. It is reconfigurable because it is software-heavy. So you need to take responsibility for its daily life and performance. To achieve this, your operating model must be able to keep up with product velocity and innovation frequency. The shift to agile platform engineering approaches is a daunting task that will be covered in more detail in Chapter 8. Modular product design, where devices or services can be quickly reconfigured, offers an adequate approach here. Otherwise it will be very hard to achieve the demanded agility and flexibility. Think of the streaming service Spotify. It is software only, but the company's agile or even tribal model only works because of a modular design.

Takeaways

1   Users are expecting to access as required rather than own a product, which leads to consuming the product as a service.

2   The software industry has proven that this shift can create enormous value. It is leading the way for the more hardware-centric industries such as automotive, industrial equipment or A&D, for example.

3   This transition is far from easy, and it requires major changes to operating models, product innovation processes, platforms and culture, as well as a complete re-architecting of the products.

6

Big Shift Three: From product to platform

CHAPTER SUMMARY

The concept of a platform is nothing new, but in the digital era of smart connected products every hardware-centric company needs to urgently work out a platform strategy. While platform-based business models offer tremendous value creation potential, not all traditional product companies need to, nor will be able to make this pivot. Several Internet and software companies have already created successful connected product platforms and many traditional hardware makers will need to consider whether to plug into these platforms as partners or compete directly.

Big Shift Three: From product to platform

CHAPTER SUMMARY

The enormous expansion of functionality injected into products by software and artificial intelligence (AI) opens new horizons. Smartened-up products, as well as becoming collaborative and responsive, can be turned into platforms. This shift from mere passive product to smart product to smart platform represents remarkable tectonic shifts in both device and software manufacture.

One of the main questions is, under which circumstances can and should a smart connected product become a platform – the centre of an ecosystem providing business opportunities for many partners?

An often-cited example is the smartphone, a platform on which third-party developers create apps, which earn them and the platform provider money. But it is important to stress that this is just one example. The world of platforms is highly complex and not all platforms are created equal. Not every hardware product can turn into a platform, and many platforms are exclusively software-based. So, there is a broad set of archetypes we must analyse to better understand how to make the most of this novel business approach.

Platforms: the new dominant driver of value creation

Over the last decade, the Internet platform giants – Amazon, Google, Apple, Microsoft, Alibaba, Tencent, Facebook and others – have achieved previously unheard of market capitalizations. In September 2018, seven of the world's top 10 companies measured by market value were Internet platform businesses.

Another wave of Internet firms also use a platform-centric business model that is built on software competence. This includes Uber, Airbnb, Netflix, Twitter, Grab, LINE, Pinterest and Rakuten, and this group too is driving massive value creation. Airbnb has received value estimates north of $30 billion[1] and video-streaming platform Netflix well above $150 billion.[2]

Figure 6.1 shows how dramatically the line-up of the biggest market cap businesses has changed in only a decade. The harsh reality is that these relatively young, 'born digital' platform leaders are often gaining value and investor attention at the expense of some incumbent product-making business.

Initially, Internet platforms posed the biggest risk to the communications, media and traditional computing technology companies. Since 2010, Internet platforms' share of total market value for this set of industries has risen dramatically from 9 per cent to 30 per cent. The largest value loss was from the communications industry, which declined

Figure 6.1   **The rise of the Internet platforms**

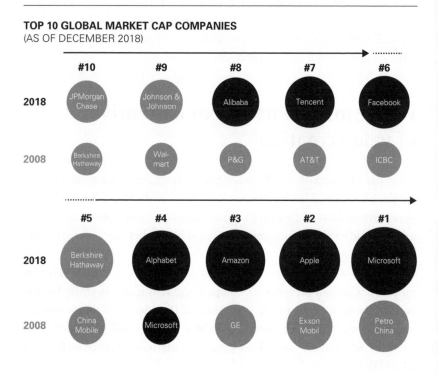

TOP 10 GLOBAL MARKET CAP COMPANIES
(AS OF DECEMBER 2018)

from 30 per cent to 16 per cent, and the computing hardware players, which declined from 23 per cent to 18 per cent of total value. The traditional software industry held constant at 13 per cent.[3]

Interestingly, the semiconductor industry has fared much better over this period than other product companies, with value share increasing as illustrated in Figure 6.2, which segments value changes within information technology-related sectors. Semiconductor companies, the makers of chips and processors, are benefitting greatly from the rise of the Internet of Things (IoT) as more and more devices are connected and each becomes more intelligent. Increasingly powerful chips are required globally to support this. Moreover, fundamental digital innovation in areas like AI, blockchain, and augmented reality is happening at semiconductor level, boding well for this industry's long-term success.

Figure 6.2 Value migration in communication & media and technology

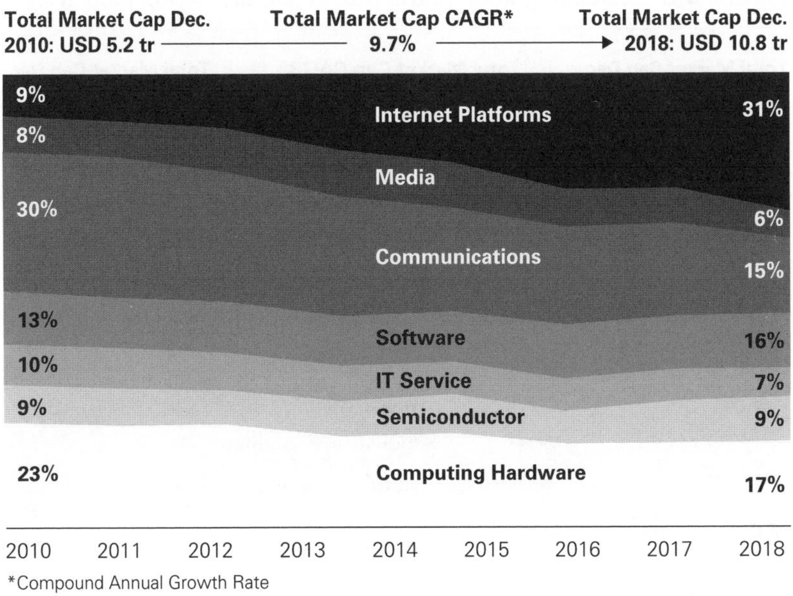
Total Market Cap Dec. 2010: USD 5.2 tr
Total Market Cap CAGR* 9.7%
Total Market Cap Dec. 2018: USD 10.8 tr
9%
Internet Platforms
31%
8%
Media
6%
30%
Communications
15%
13%
Software
16%
10%
IT Service
7%
9%
Semiconductor
9%
23%
Computing Hardware
17%
2010
2011
2012
2013
2014
2015
2016
2017
2018
*Compound Annual Growth Rate

Now let's look at the impact on industrial and consumer goods companies. Since 2010, our research shows, Internet platforms' share of total market value versus these sectors has risen even more dramatically from 12 per cent to 38 per cent. During the same period, the value share for hardware-centric companies such as industrial equipment manufacturers more than halved, while those of heavy equipment makers and producers of cars and durable consumer goods also shrank substantially.

The very real risk for traditional product makers in various industry sectors is that they will be squeezed between the expanding value of platform companies and the rising value of semiconductor companies, as both these industries continue to drive fundamental innovation. There is a direct analogy to the personal computer (PC) industry, which evolved to the point where the vast majority of its profits were controlled by two enterprises, one a software company called Microsoft and the other a semiconductor business called Intel.

Figure 6.3   **Value migration in traditional manufacturing industries**

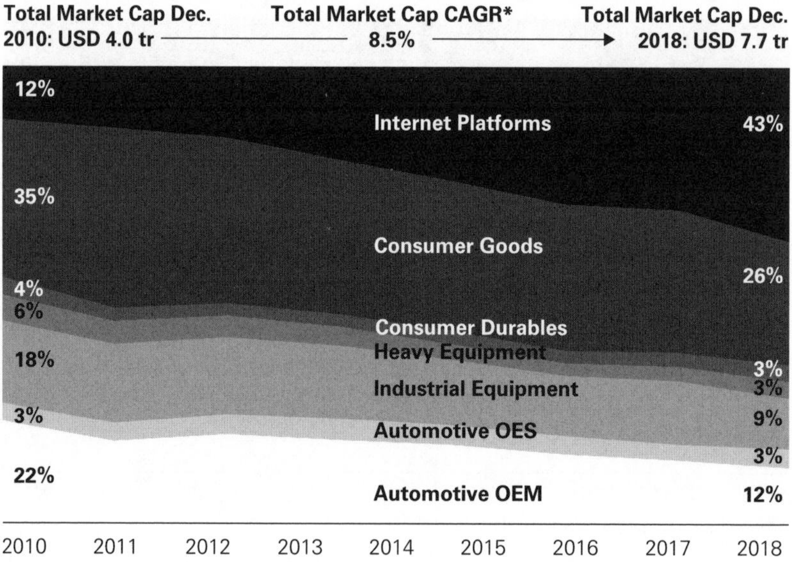

Total Market Cap Dec. 2010: USD 4.0 tr	Total Market Cap CAGR* 8.5%	Total Market Cap Dec. 2018: USD 7.7 tr

12%	Internet Platforms	43%
35%	Consumer Goods	26%
4%	Consumer Durables	
6%	Heavy Equipment	3%
18%	Industrial Equipment	3%
3%	Automotive OES	9%
		3%
22%	Automotive OEM	12%

2010   2011   2012   2013   2014   2015   2016   2017   2018

*Compound Annual Growth Rate

The only hardware product company in our global top 10 market capitalization list is Apple. But this manufacturer of upmarket digital consumer devices such as smartphones, tablets, voice assistants, laptops and desktops has led the way in the creation of platforms around products, making it really a hybrid: a product-plus-platform company. It also designs most of its core semiconductors itself, so is arguably a product-plus-platform-plus-semiconductor business. We will explore how Apple attained its position later in this chapter.

Apple's hardware represents engineering of the highest quality and advancement, differentiating it from more mundane devices for instance in the smartphone market. This matters because the high quality helps Apple continue to be a leader in smartphone product sales, which in turn enables it to attain and then maintain a platform status that rivals find hard to disrupt. So while software and digital intelligence components provide the crucial value weight in smart and connected products, high-quality and high-performance hardware engineering is still critical. This should be a guideline for all hardware makers facing the challenge of turning their hardware products into platforms.

Carmaker Tesla can serve as a good role model here, where vehicles, not unlike Apple's high-end smartphones, are designed to work as platforms. An executive from Tesla's Connected System Engineering unit told us:

> 'We have to think of these vehicles as sensor platforms that experience and data-record the road life around them via cameras, ultrasonic sensors and radar. That data can be used to inform the development of driver assistant and autopilot capabilities or to develop driver-training sets.'[4]

Many types of platform, not all are alike

Simply put, while a product is something you sell, a platform is infrastructure that enables interactions between various actors.

All successful platforms create what are called network effects where the value of the platform increases as the number of users and the usage increases. Economists call this 'demand-side economies of scale', which is very different from the traditional 'supply-side economies of scale' that govern most hardware-based products companies.

In the supply-side version, companies focus on building scale in areas like manufacturing, supply chain, procurement, and sales distribution. It's cheaper per bicycle, say, once you have the infrastructure, to produce a million bicycles than just one. Again, the value is also increased by the number of users, but here the users are not networked, they only interact with the selling company and it's a one-off: they buy a bicycle.

By contrast, network-based platform companies focus much more on scaling the number of users and usage of the platform. This model often requires large upfront investments to build out the capabilities and the user base until a tipping point is reached where the cumulative value of the services provided by the platform is greater than the cost of developing and running it. Once the platform successfully passes this point, the value creation potential is enormous.

The initial network-based platforms were physical systems like roads, utilities and railroads, followed by phones and now the Internet. There are now platform businesses active in transactions, data, social communities, development and potentially in AI.

Figure 6.4 (see page 113) ranks the major five platform types we can identify today in business practice.

Nine key components of a successful platform

While the mix of success factors for a platform varies based on the type of platform, the common characteristics of winning platforms typically include the following nine features:

Figure 6.4 The five major platform types

	Business Model	Short Description
Marketplace Platforms	Orchestration Matching	Facilitate selling and buying of goods and services between sellers, distributors and end customers.
Live Examples	eCommerce: Amazon, Alibaba, eBay Hospitality: Booking.com, Trivago, Travelocity, Hotels.com, Kayak On Demand Content: Netflix, HBO Now, Spotify Payment: PayPal, Apple Pay, Amazon Pay, Google Wallet	
Social & Collaboration Platforms	Collaboration	Enable developing connections, enhance communications, speed networking and share data between its users.
Live Examples	Social Media: Facebook, Twitter, Pinterest, Instagram, LinkedIn Instant Messaging: WhatsApp, Skype, LINE, WeChat, Business communities: Skype for Business	
Sharing Platforms	Matching	Leveraging the power of sharing economy and shared data.
Live Examples	Shared Data: Waze, Flixster, Yelp!, Zillow Sharing Economy: Crowdfunding (Kickstarter, Indiegogo), Place Renting (Airbnb, Couchsurfing), Ride/Car-Sharing (Uber, Lyft, Zipcar, DriveNow, Car2go, Blablacar)	
IoT Platforms	Creation Orchestration	Bring together information from sensors, devices, networks and software that work together to unlock the value in data.
Live Examples	B2B: IBM's Watson IoT platform, GE Predix, Schneider Electric EcoStruxure, Siemens MindSphere, Philips Interact IoT B2C: Haier U+ Platform, Apple HomeKit, Google Smart Home Platform	
Developer Platforms	Creation	An infrastructure or set of technologies connecting end users and developers. The platform serves as a base on which other applications, products or processes are developed.
Live Examples	Cloud Platforms: AWS; Google Cloud Platform, Microsoft Azure Operating Systems: Mobile (Apple iOS, Google Android, BlackBerry OS, Windows Mobile), Desktop (Windows, Mac OS, Linux) Browsers: Google Chrome, Microsoft Edge, Mozilla Firefox	

Figure 6.5 Key components to a successful platform

1 **Value proposition based on network effects.** A platform's value increases with the number of users and frequency of usage. This creates a circular, self-reinforcing effect. As most platforms have two sides – for example buyers and sellers or users and developers – it is critical to develop a compelling value proposition for each platform stakeholder. A very good example is Waze. Launched in 2007, this is a traffic and navigation platform using crowd-sourced driver data shared by users to benefit the Waze community and improve driving quality. Today Waze has become a mass application with 100 million active monthly users in 185 countries.[5]

2 **Compelling user experience.** At the core of all the leading platforms today is an easy-to-understand and easy-to-use experience. Think of the simplicity of using the Google search engine, ordering a cab on Uber (or Didi in China) or buying an app from the App Store or Google Play.

3 Strong ecosystem. Leading platform providers have thousands or even millions of developers, innovation partners, system integrators and service partners supporting their platform. This requires strong ecosystem support, including solution developer kits (SDKs), application program interfaces (APIs) and developer support as well as ecosystem development activities. In the case of industrial businesses, platform partners can even comprise other product makers or third-party service providers. We will touch on a few examples further down. In a healthy ecosystem, all participants co-exist and share value continuously. Platforms offer various transparent value-sharing mechanisms for partners but the decision has to be taken by the platform initiator on whether the platform is designed as an open widely embedded co-creation platform or more as a closed platform with stricter control by the platform orchestrator. 'Creating a winning platform is a very difficult challenge. You need to focus on user value and develop as fast as possible. Use whatever you can to push the adoption among prospective platform users to then enable you to have the support for an ecosystem', says Steve Myers, CEO of Mindtribe.

SMART VOICES **Q&A with Professor Michael G Jacobides, Sir Donald Gordon Chair for Entrepreneurship & Innovation, London Business School**

How do you assess the relationship between ecosystems and platforms in the context of smart connected products?

As it looks, platforms, such as the operating systems of smartphones or a mobile phone bandwidth standard, are developed more often than ecosystems. Sometimes platforms do not spawn ecosystems, as they have no concomitance between their members. To me the defining difference between a platform and an ecosystem is that in an ecosystem you care about the development of things that exist in the rest of the system, whereas in platforms you might not.

In that regard, ecosystems create some more specific and non-generic relationships, which essentially means that you have a slightly more closed system that sets itself against other closed systems.

Does that mean platforms are less strategically relevant for smart connected products?

No, I still think that platforms are important. The fact that they are based on much looser links can end up with them becoming generic and possibly regulation will push them to become even standardized. If these become standard for every user – an example is the 5G wireless standard – I do not think that those platforms are strategically and economically meaningful anymore. But be it as it may. I think smart products require interconnections to work. The interconnections they get are either generic and provided by platforms or they are going to be more specific, meaning that they will connect in a more seamless way with other providers that integrate them and affect either the corporate value or the customers' quality of life. The latter ones, I believe, will make the biggest impact commercially. There is a clear connection in my mind between the smart product and the platform and ecosystem dynamics, as well as the strategic issues that firms have to face as they compete with other platforms and with ecosystems.

4 **Business model clarity.** The way a platform owner makes money is often separate from and sometimes almost completely divorced from direct usage of the platform itself. For example, Google and Facebook do not charge consumers to use their platform but make money primarily from advertisers targeting ads based on users' search queries. Other revenue sources can include transaction fees, usage fees, or licensing rights. Fascinatingly, the top platform players today all have very different business models and business mixes. There is no single winning formula, but clarity of the business model is essential. See Figure 6.6 for a breakdown of business divisions among Internet platform leaders.

Figure 6.6   Platform titans have diverse business models

SHARE OF 2017 REVENUE BY BUSINESS MODEL

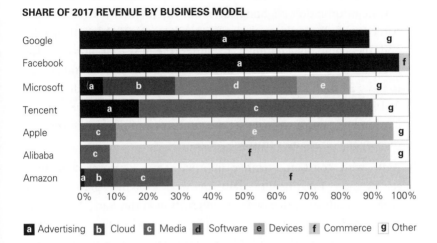

a Advertising b Cloud c Media d Software e Devices f Commerce g Other

5 **Differentiated data and content.** For platform models based on data, it is essential to create or aggregate proprietary data sets. This can be user-generated data in the case of consumer review sites like Yelp!, aggregated public data in the case of real-estate sites like Zillow, or usage data such as in the case of ComScore.

6 **Marketplaces.** While not all platforms contain marketplace functions, those that do must achieve the critical mass tipping point of both buyers and sellers. But the term marketplace also applies here in a wider sense, as a hub connecting people and enterprises in a business-to-business environment. Once such a marketplace is established in a particular customer segment or industry it creates tremendous network effects. Just think of the power of the App Store or of Google Play. There are dozens of smartphone manufacturers today, but basically only these two app marketplaces.

7 **Scalable.** Ability to scale to millions and even billions of users globally. This can only be accomplished with state-of-the-art infrastructure that is typically cloud-based and open. Very few traditional product companies have so far embraced this infrastructure model.

8 **Digital trust and security.** This issue is becoming increasingly important given the recent, very high-profile security breaches and abuses of private information blighting several of the Internet platform giants. After Equifax's data breach, its CEO, CIO, and CSO lost their jobs, and the company continues to suffer the financial impact.[6] The personal data of 145 million eBay customers was stolen.[7] Yahoo had to pay $35 million to settle charges for failing to properly disclose its 2014 data breach affecting 500 million users.[8] All leaders are working hard to build much stronger security features into their platforms and thereby maximize user trust.

9 **Agile with constant, iterative improvements.** Platform companies release new functionality daily or weekly. While platforms are typically built to enable one core interaction at the start, the winners add more features very quickly as they evolve. Most of these features are discovered as the platform is consumed in ways not originally intended. Traditional product companies have product development cycles typically measured in months and even years, and generally only focus on improvement to the core product. Speed of innovation is a major competitive weapon, and most product companies are losing this battle. One simple example is Tesla versus traditional cars. Tesla absolutely views itself as a platform company and regularly publishes software updates that fundamentally upgrade the car. One of our interviewees says, 'The same Tesla car a customer acquired in 2013 is today a much better car than it was when it was bought – due to permanent software updates.'[9]

The network effect is the decisive trait of winning platforms. Network effects create a circular, self-propelling pull momentum when providers and users generate mutual network value for each other, resulting in two-sided benefits. The two sides do much of the work required to grow the platform and the fortunes of the company behind it. This stands in stark contrast to the linear value chains of traditional product companies, which require constant investment in marketing and sales to push their products in a crowded marketplace. And this has ramifications for branding, as it has to be decided which branding budgets, if any at all, are dedicated to which channels within a platform.

The imperative of platforms for product makers

As agile platform pioneers continue to rapidly grow, many traditional product companies have been too slow to both realize the importance of platforms and develop a clear and compelling platform strategy. But you don't just need a good strategy – and management backing – you need the means to implement it. Even those with a clear vision have struggled to develop the requisite skills and capabilities, and execute at the pace needed to compete in the digital era. Missing this generational shift or getting it wrong could result in a long stagnation or decay.

SMART VOICES   **Q&A with James E Heppelmann, CEO, PTC**

Could industrial companies not develop such software platforms themselves?

Manufacturers might well have originally thought to develop such technology. But I think, given the speed at which everything is moving, and all the new use cases that are coming out, people realize that this would be a tremendous effort for only one end user benefitting from it. We also think that an IoT platform for the smart connected product is the classic case of the 'long tail' problem. Twitter is one application used by millions of people. If somebody is trying to do smart connected products they start with one application to monitor the fleet. Pretty soon they need another application to try to do analytics and predict downtimes and then they might need another application that would inform sales and account managers and yet another application might inform them what might happen on the customer side. So we talk about a whole string of applications that might be used by very narrow audiences but would create tremendous value. A universally adaptable platform is the right answer to that in my view. And industrial clients seem to have taken that on board – after all we are not the only one in this IoT platform market.

To increase the sense of urgency further, many of the newer digital leaders who started as pure Internet or software-based platforms are increasingly moving into the hardware business to take full advantage of the opportunities they see traditional companies overlooking. Google, Amazon, Facebook and Microsoft all now have significant product divisions. Even some semiconductor companies like Intel are moving up the value chain to offer complete reference designs and, in some cases, branded products.

These Internet platform leaders have also moved, or are actively considering moving into more traditional industries on the back of the abundant financial resources they have amassed from their platform businesses. Think of the numerous tech companies involved in autonomous vehicle projects, trying to eat the car manufacturers' lunch. Their advanced operating models, vast customer bases, and low marginal expansion costs mean these rivals can quickly become established, change the rules of the game, and undermine incumbents.

This kind of disruptive environment is also an opportunity for entirely new entrants. Think of how businesses like NIO, Byton and even Dyson are entering the automotive sector.[10] Even former suppliers can begin snapping at the heels of their long-term clients, as in the case of car interior specialist Faurecia.

A Tesla executive from the carmaker's connected systems engineering unit told us, 'Smart connected products have all platform character in my view. In many cases they are not one single platform but a portfolio of platforms.'[11] While a platform strategy is essential for product companies, not all product companies can or should directly develop platforms. Many will choose to partner with new or existing platform leaders.

Winners will master the strategic use of digital technologies to build successful platform business models. Losers will be missing out on a goldmine. Platform-driven interactions are expected to create approximately two-thirds of the $100 trillion value at stake from digitization by 2025.[12]

The product makers finding the way

Let's come back to Apple as the leading role model for traditional hardware companies thinking of getting into platforms. Once merely acomputer hardware maker, it has transitioned to become a platform company built around smart connected products.

Indeed, Apple today has not one but several successful platforms. Its App Store is a marketplace connecting users with developers' applications; Apple Pay is a payment platform connecting merchants with consumers; and Apple Maps is a data-based platform that aggregates massive global geospatial data for consumer use. In fact, iOS – Apple's operating system – is also a development platform that developers use to create mobile applications.

The App Store's platform success is especially phenomenal. Launched in 2008, it now includes an ecosystem of over 20 million registered developers[13] who have created 2.2 million applications.[14] These apps, each creating value for their makers and Apple, have been downloaded more than 140 billion times.[15] Based on Apple's 70/30 split with developers, the App Store has so far generated $38 billion for the company.[16]

Note that the company continues to generate the majority of its revenue and profits from selling hardware: iPhones, iPads and Mac computers.

Following Apple's lead, a string of incumbent industrial firms have already made significant steps into similar business models. Following are six prominent examples.

Ford, in a joint effort with Autonomic, is building the Transport Mobility Cloud, a new open platform connecting and coordinating mobility services such as cars, bike-sharing networks, buses, trains, and ride-hailing services within a city in real time.

GE Predix is an IIoT platform, a platform as a service, for gathering, storing, and analysing industrial machinery and equipment data. GE aims to make it the industry standard for connecting machines and equipment, similar to Apple's iOS or Google's Android. The Predix Platform provides an industrial-grade analytics library and framework to create machine learning analytics tailored to digital twins and applications, and other manufacturers are keen to follow the same path. When applied to IoT data streams, edge and cloud-based analytics can detect anomalies, direct prescriptive controls, signal predictive maintenance alerts, and more.

Schneider EcoStruxure Power is an open, IoT-enabled, interoperable architecture for clients using low- and medium-voltage equipment. It leverages advanced technology in IoT, mobility, sensors, cloud computing and cyber security and has been deployed in over 480,000 installations, with the support of over 20,000 system integrators, connecting over 1.5 million assets.[17]

John Deere has developed MyJohnDeere, an open information platform supplying agricultural producers with critical information such as production or farm operations data. Data is collected through sensors from machines, operators and dealers connected to John Deere equipment and is shared across those same users.

Faurecia, building on the expertise of Parrot Automotive and Accenture, is developing an open platform to manage all car cockpit functions through data analytics, AI, telematics and cloud services. Faurecia has already received €1.5 billion in orders from customers for its new Smart Life on Board technologies. By 2025, these sales will reach €4.2 billion, showing an average annual growth between 2020 and 2025 of 33 per cent.[18]

Haier, the Chinese domestic appliances giant, has created Cosmoplat, an open marketplace platform allowing consumers to order washing machines, dishwashers or refrigerators according to their individualized specifications directly from a dense network of hardware producers.

Haier's role here changes from product developer and manufacturer to ecosystem orchestrator, curating, connecting, coordinating, integrating, and managing different stakeholders.

These examples demonstrate the feasibility of the shift from traditional product to product-plus-platform. It is, no doubt, more difficult when you are a manufacturer with 150 years of engineering under your belt, not a youngish software-driven disruptor, but you can do it, and your smart connected product should feature centrally in your platform approach.

Internet platform giants: friend and foe?

Market researcher IDC predicted that more than 50 per cent of large enterprises – and more than 80 per cent of enterprises with advanced digital transformation strategies – would have created or partnered with platforms in some form by the end of 2018.[19] It is important for traditional product businesses to treat platforms, especially those provided by the Internet giants, as potential trusted partners, but also pragmatic competitors. In some cases they will strongly compete or even try to disrupt your business. In others they will act as partners or facilitators. You might simply use the Apple App Store or Google Play as a channel to distribute application software for your own hardware product, entirely to your company's benefit.

Our view is that all product-based companies need to have a clear friend-or-foe strategy when it comes to defining their relationship to the Internet platform titans and other rivals. You need to figure out whether to build a platform, and if not, where and when to buy into or partner with one.

For many product companies, choosing to partner will be the logical answer as it would be virtually impossible to either catch up with the leaders technologically or match their investments, which are already astronomical and mostly increasing. The world's largest R&D spender is now Amazon. Alphabet, Google's parent, is number two, and Microsoft is number five as Figure 6.7 shows. In fact, seven of the top 10 R&D spenders globally are now either Internet platforms or high-tech leaders.

These players are investing massively, both for today's needs and for future battlegrounds, including autonomous vehicles, augmented reality,

Figure 6.7   Top 10 global R&D spend leaders

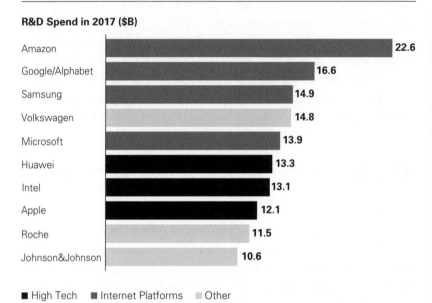

R&D Spend in 2017 ($B)

Amazon	22.6
Google/Alphabet	16.6
Samsung	14.9
Volkswagen	14.8
Microsoft	13.9
Huawei	13.3
Intel	13.1
Apple	12.1
Roche	11.5
Johnson&Johnson	10.6

■ High Tech   ■ Internet Platforms   ▨ Other

SOURCE Based on: Fox, J (2018) Amazon, the Biggest R&D Spender, Does Not Believe in R&D, Bloomberg, 12 April [online] https://www.bloomberg.com/opinion/articles/2018-04-12/amazon-doesn-t-believe-in-research-and-development-spending [accessed 6 December 2018]

mapping and geospatial data, AI and voice assistants, and payment platforms.

So, the question of whether to compete or partner must be taken quite seriously and, in our opinion, cannot be answered in the abstract. Product players need to consider the specific types of platforms to invest in. Take another look at our table of platform types in Figure 6.8 for some more granular insights into the possibilities for incumbent industrial firms.

Different businesses have positioned themselves differently in the friend-or-foe decision. Automotive players such as Volkswagen or BMW invest a great deal in data technology and data processing centres and even quantum computing projects. German technology behemoth Bosch has thrown the gauntlet to Google by pushing massively into data technology

Figure 6.8   Competition opportunities per platform type

	Potential for Product Manufacturers	Rationale
Marketplace Platforms	Medium	Well-established consumer marketplaces with limited ability for new entrants.
Examples COSMOPlat Developed by Haier Amazon Pay Apple Pay eBL Blockchain Platform		Product companies can leverage their customer base to create micro-marketplaces, particularly in the B2B world.
Social & Collaboration Platforms	Low to Medium	Mature social area with limited potential.
Examples My JohnDeere		Opportunity to create industry-specific or customer segment-specific 'micro-communities'. Note: Can be very helpful to strengthen product differentiation, but difficult to monetize given smaller user base.
Sharing Platforms	High	More fragmented market globally, with higher new entrant opportunity.
Examples GM Maven Ford Mobility		Look to leverage installed base data to create data platforms and/or create industry consortia to create data platforms, including leveraging Blockchain.
IoT Platforms	High	Leverage the power of installed base of products.
Examples GE Predix Philips Interact CAT Connect Smart Home Platforms		
Developer Platforms	Low	Very difficult to create a broad new developer platform, but possible to create a 'sub-platform' that is built on top of current leaders to allow third-party developers to access and customize products.
Examples Signify (Philips) Hue built on AWS Fanuc Robots Faurecia CIP		

around smart home solutions. Through a new smart home subsidiary, Bosch will deal directly with the end consumer for the first time.

Other businesses go down the 'friend' path and become allies of the big Internet giants. Volvo has been a pioneer in marrying digital technology and automobiles. It has turned to outside providers like Ericsson, a Swedish maker of telecommunications equipment, for computer technology. It has announced it will install Google's Android operating system in new cars from the beginning of 2019 and cooperates with Uber to develop self-driving cars.

Finally, note that adopting platform models does not mean giving up on existing product-based business models. In fact, the existing business lines will often provide the new platform's foundational strength – and in many cases the funding for establishing a platform in the first place. At its core, Signify (formerly Philips Lighting) is still in the business of making lighting, HP will still make printers, Boeing aeroplanes, Michelin tyres and Mercedes cars. But the platform business model offers such companies the opportunity to leverage these products in the digital era in innovative and highly value-creating ways.

Takeaways

1 Platform business models are creating enormous market value.

2 Every product company must have a platform strategy, and determine whether to build their own platform or partner as well as what type of platform(s) models to participate in. Ignoring is not an option.

3 Many product companies will choose to partner with today's Internet giants who are platform leaders, but all need to understand the risk versus reward of their choices.

7

Big Shift Four: From mechatronics to artificial intelligence (AI)

CHAPTER SUMMARY

This chapter highlights the enormous significance of AI in accelerating the trend towards smart connected products. Products, born as the products of mechanical engineering, are already now significantly dominated by software and will soon be transformed by AI. To software-enable a product does not yet mean to render it smart in the sense of truly autonomous thinking and decision making. Only AI can do this and its broad introduction to products amounts to a quantum leap in innovation approaches, product adaptability and responsiveness. In the following pages, we hammer out these crucial differences and sketch out what AI can mean for connected products when delivering services.

The rise of software and then software-enabled AI technologies represents a change for product-making companies at least as significant as the change from steam to electrical power. Even modern computers have seen some mechanical elements replaced by software-laden hardware in a very short time, as in the case of hard-disk technology giving way to mechanics-free flash storage technology and now to cloud-based storage as a service.

Now the change is affecting all sorts of traditional industrial products. For a century, cars consisted of a mix of combustion engines and various other engineered parts combined with growing amounts of electronics and microelectronics. Today software has come to dominate the automotive works. Coded program lines run the engine ignition, supervise vehicle safety, provide entertainment to passengers and take over navigation.

Today's leading product and service providers have built significant software and digital capabilities, and those that have not are under serious threat of disruption. The fastest-growing entertainment providers are software-based Internet platforms: Netflix, Spotify, Apple iTunes. The natural resources sector has become a major provider of data visualization software. Retailers and logistics businesses have essentially turned into software backbones, steering fleets of trucks and stores. Telecommunication has been disrupted by software alternatives such as Skype or Zoom. Recruiting, advertising and finance are reaching ever-deeper levels of software saturation.

More broadly, what has happened is this: over decades, in both hardware products and manufacturing methods, electronic, micro-electronic and digital technologies have grown in importance, laying the foundations for increasing levels of software. This in turn has created a bedrock for artificial intelligence, or AI.

AI: a quantum leap for product companies

Even compared with all the other hyper-speed advancements of the software age in the product world, we believe that AI represents a quantum leap. A matured AI technology stack means extremely advanced levels of software intelligence, enabling products to live an autonomous life. This is already underway and is changing how products are conceived by their makers, how they are built, how users interact with them, what character traits are possible in a product, and even the fundamental definition of what a product is.

There is a lot of confusion among executives about what exactly AI is and is not, so let's take a moment to explain the basics. First and foremost, AI is not one singular technology but a collection of them, some of which have been around for decades and some of which have broken through quite recently.

More precisely, it is a combination of information technology systems, tools and methodologies such as data mining, pattern recognition, visual sensing and natural language processing, that enables machines to sense, comprehend, act and learn on their own or with minimal human augmentation. The ability for an AI-powered product to autonomously learn, adapt and improve at increasingly lower cost is a game-changing development.

Figure 7.1 lists the foundational AI technologies, grouped around their ability to support key product functions.

As you can see, there are four humanoid AI capabilities that businesses can now inject into products.

First, AI allows products to sense. Products can now actively perceive the world via processing text, images, sounds and speech. Voice and face recognition are already widely used in AI systems, mimicking human hearing and sight to take in environmental information for interpretation. These new capabilities complement older sensor technology systems, collecting the physical data vital to smart products' operations.

The second, perhaps more important, capability is that AI can make products comprehend in a humanoid way. Natural language processing, inference technology and knowledge representation enable AI systems to analyse and semantically understand the information collected from outside by adding meaning and insights. This forms the basis for autonomous decision making.

Figure 7.1   Foundational AI technologies

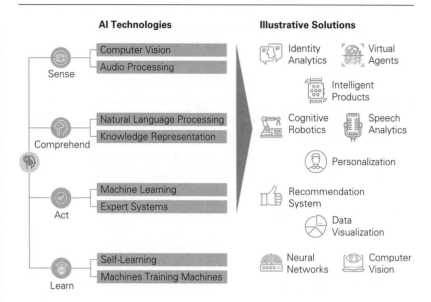

Third, the product can increasingly act and make decisions. A good example of this is a self-driving, autonomous car.

And finally, perhaps even more crucially, AI systems can learn, their incessantly growing knowledge stocks also channelling their understanding into measured and rational interventions in the surrounding world. While today, AI engines need training conducted by humans, increasingly the products will become self-learning and self-correcting.

These fundamental traits open entirely new horizons. AI-enabled products can now, all of a sudden, deliver hyper-personalized outcomes to users at unheard of levels of speed, accuracy and convenience.

The different AI technologies are often combined to create compelling user solutions, such as virtual agents or voice assistants like Amazon Alexa or Microsoft's Cortana.

Such assistants require Natural Language Processing (NLP) technology to understand or comprehend human speech. This is combined with machine learning (ML) to analyse human questions and then give appropriate answers. Everybody in the AI scene – and many more beyond it – knows about Google's impressive recent presentation of its AI algorithm, Duplex. It rang a real restaurant and effortlessly conversed with a waitress, discussing table reservation options in fluid human language.

Figure 7.2 shows a schematic of how such assistant systems work. According to Danny Lange, the former head of machine learning at Uber, AI has finally broken out of the research lab and is fast becoming the cornerstone of business disruption.[1] In the smart hardware world, it primarily transforms the user interface and user interaction with the product, and enables high-value services around the product. Today, AI is already deployed to various degrees in many hardware industries and its role in the daily lives of both consumer and business users is set to

Figure 7.2   Technology components of a virtual customer assistant

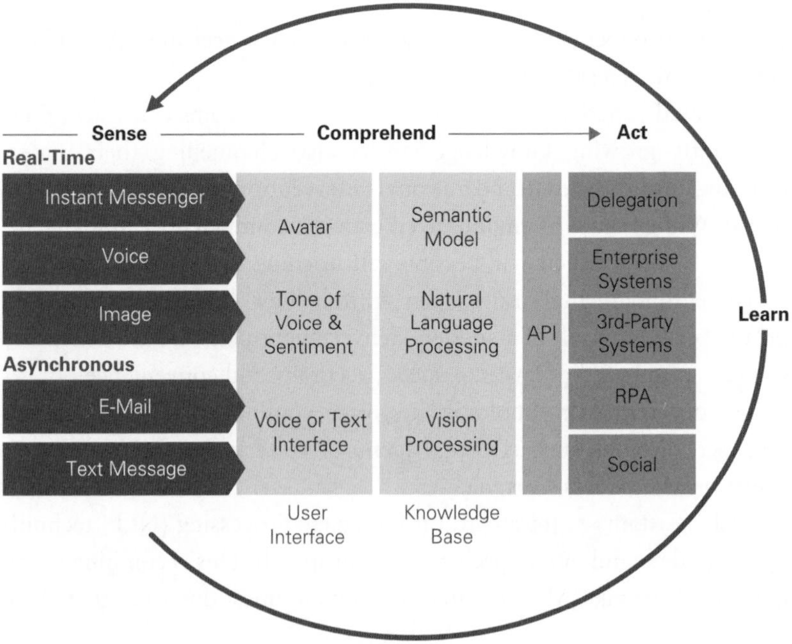

grow fast. Let's take a quick, kaleidoscopic overview of how AI is already in daily use in both manufacturing and products:

1 Industrial assembly lines can be reconfigured through AI. Engineers at the Fraunhofer Institute of Material Flow and Logistics (IML) have been testing embedded sensors to create self-adapting assembly lines in car plants. Essentially, the line itself can modify the steps in its process to fit the demands of various features and add-ons for highly customized cars. Thus, instead of engineers designing an assembly line to make one car model series at a time, these lines can adapt as needed down to a lot of one.[2]

2 AI-steered robotic arms are used in factories to apply hot glue to widgets, install windshields, and smooth jagged metal edges, among other tasks. Also, traditionally, when a robot's job changed, engineers had to reprogram them. In contrast, the new robotic arms developed by Japan-based robot specialist Fanuc, in partnership with Japanese software maker Preferred Networks, adapt on their own. To do this they use an AI technique called deep reinforcement learning, in which the robot is given a picture of the successful outcome and then uses trial and error to figure out its own solution. According to Shohei Hido, VP of Research at Preferred Networks, the arm takes eight hours to become at least 90 per cent accurate at a new task – virtually the same time and accuracy achieved through programming. With the arm now autodidactic, the human expert is freed to do further complex tasks, especially those requiring human judgement. What's more, once a robot learns a task, it can share its knowledge with other robots in the network. Eight arms working together for an hour can learn as much as one working alone for eight hours. Hido, who calls this process 'distributed learning', says, 'You can imagine hundreds of factory robots sharing information.' Carmakers use this approach to educate algorithms that are meant to steer vehicles autonomously.[3]

3 An example from healthcare: now that so much data about patients' individual genomes and responsiveness to various chemical compounds is available, it simply doesn't make sense to deploy one-size-fits-all treatments. AI is now enabling the era of 'personalized medicine' based on genetic testing. In the past it was virtually impossible to analyse and manage all possible treatment combinations for each patient. Today, intelligent systems are taking over the job. Decades from now, perhaps even sooner, it will seem absurd that doctors prescribed the same treatment to a wide range of patients. Everyone's treatment will be personalized. Along these lines, analytics firm GNS has been crunching huge data sets to match specific drugs and nondrug interventions to individual patients. The company can thus improve outcomes and lower costs to save hundreds of billions of dollars, according to co-founder Colin Hill. Individualized treatments could solve an especially critical problem in clinical trials: more than 80 per cent fail due to some level of mismatch between patient and drug.[4]

4 The automotive sector and its suppliers are working towards AI technologies fulfilling elaborate roles in vehicles. On-board sensors pick up a wide variety of data both about driver and driving conditions. AI could, for example, square data on your individual driving behaviour, tiredness levels and features usage habits with ad hoc measured situational data such as weather or traffic parameters. AI-enabled cars could then reconfigure their whole character by changing user interface designs to activate or deactivate certain features.

Three factors enabling the rise of AI

Three achievements have allowed AI to become central to the future of smart connected products: 1) increasing compute power and data storage capacity, both on the device and in the cloud; 2) big data analytics; 3) the rise of human and machine tools and a workforce capable of linking the technology to solve business problems.

Cloud and data storage are now available in quasi-unlimited quantities and no longer pose any obstacle for the quick spread of data-rich AI technology. Public cloud computing is estimated to reach a value of almost US \$302 billion by 2021 worldwide, and data storage capacity has also become almost ubiquitous.[5] According to IDC, worldwide IT cloud services revenue is estimated to grow from $103 billion in 2016 to $277 billion in 2021, while the cloud storage market is projected to witness a compound annual growth rate of 29 per cent to reach a total market size of $92.5 billion by 2022, up from $25.2 billion in 2017.[6]

The second trend, big data analytics, is now performing at industrial scales on the back of ever more powerful processors and storage technologies. Global data volume will grow 10-fold by 2025 and reach 163 zetabyte as more devices have become connected.[7]

AI feeds on virtually any data, be it images, text or audio, as well as structured and unstructured data. Some scholars therefore see data as being the equivalent for AI as food is for humans. Mass data management and turbo-charged computing speed create a world in which there are simply vastly more inputs for AI to intelligently respond to and learn from and thereby deliver ever more personalized outcomes.

Long-held paradigms of economic theory are starting to shift under the rapid spread of AI. Traditionally labour, capital, land, intellectual property and physical assets are considered the five principal economic production factors. Sometimes referred to as the 'third workforce', AI is really a hybrid between human labour and capital assets, but is seen by many economists as a sixth economic production factor because of its massive potential for efficiency and productivity gains. We believe the creation of new human and machine solutions that link enabling AI technologies to people executing specific tasks or processes is the third factor driving the rapid adoption of AI.

Figure 7.3 illustrates how AI can significantly accelerate growth, profitability, and sustainability in businesses over time.

The unstoppable ascendancy of voice

Voice assistant functionality is one of the most vital technologies making up AI for product-making companies. For many products, voice is the next logical user interface (UI) replacing the keyboard or touchscreen as the main basis for interaction between user and device. Some analysts

Figure 7.3 AI can help restore and accelerate profitability

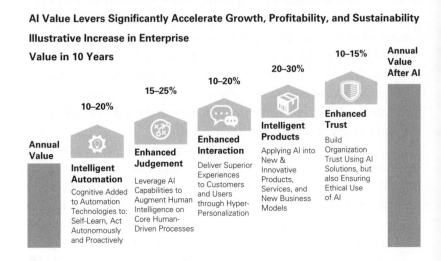

AI Value Levers Significantly Accelerate Growth, Profitability, and Sustainability

Illustrative Increase in Enterprise Value in 10 Years

Intelligent Automation
Cognitive Added to Automation Technologies to: Self-Learn, Act Autonomously and Proactively
(10–20%)

Enhanced Judgement
Leverage AI Capabilities to Augment Human Intelligence on Core Human-Driven Processes
(15–25%)

Enhanced Interaction
Deliver Superior Experiences to Customers and Users through Hyper-Personalization
(10–20%)

Intelligent Products
Applying AI into New & Innovative Products, Services, and New Business Models
(20–30%)

Enhanced Trust
Build Organization Trust Using AI Solutions, but also Ensuring Ethical Use of AI
(10–15%)

Annual Value → Annual Value After AI

estimate that by 2020 there will be over 1.6 billion users worldwide of voice assistant technology.[8]

In the last five years, usage has already become widespread, starting first on smartphones and then on smart speakers for the home such as Amazon Echo and Google Home. Consumer acceptance has been huge, sparking an industry-wide AI frenzy, a rush to embed the technology in a wide range of devices, from cars to domestic appliances.

The first voice assistant embedded in a commercial product was Siri, launched on the Apple iPhone 4S in October 2011.[9] Google followed several years later with the Google Assistant, which became widely available for Android-based phones in 2017.[10]

The smart speaker first appeared only in 2015 with the launch of Amazon Echo,[11] but three years later it is already one of the great success stories in consumer electronics history. Google Home followed a year after Echo,[12] followed by Apple's HomePod in 2018.[13] From being a non-existent category three years ago, this market has exploded and is expected to reach sales of 56 million shipped units globally in 2018.[14] These assistants have significantly enhanced these companies' ability to foster customer engagement. Market capitalization of both Amazon and Google spiralled upward after the launch of their respective AI home devices.

One factor driving the rapid growth of this product category is the dramatic expansion of global language support. By the end of 2018 Google Assistant supported 30 languages, covering 95 per cent of the global smartphone user population.[15] The quality and accuracy of assistant services is also improving at lightning speed, with the leading assistants today capable of understanding over 95 per cent of words globally, while accuracy in answering questions has now exceeded 80 per cent.[16]

At this point, the technology is, without doubt, mature enough to drive high levels of customer satisfaction and adoption. The Accenture Digital Consumer survey of 21,000 customers in 19 countries in 2018 showed strong interest in usage of, or satisfaction with 'stand-alone digital voice assistant' products. Ninety-four per cent of all consumers are satisfied or very satisfied with their voice assistant devices – an incredible rating for a new device category.[17]

Based on this success, a wide variety of product makers are now looking to embed this technology in their devices. Early examples in the consumer area are voice-enabled speakers by Sonos, Windows 10 from Microsoft, and cameras by the home technology specialist Nest. Indeed, Amazon and Google are engaged in an arms race to get their AI technology embedded into as many third-party devices as possible.

Watching a digital home assistant performing its tasks, one gets a notion of how advanced AI technology has become and what it has in store. It plays music, provides information, and delivers news and sports scores. The speaker reads you the weather report and controls your smart home by changing lights when your mood changes. It is updated through the cloud automatically and learns incessantly what it is being told or asked. The more you use the device, the more the technology adapts to your speech patterns, vocabulary and personal preferences. The next generations of these platforms will become even more impressive as they will understand and communicate back in complex sentence formats and be able to detect emotions and moods and respond accordingly. In short, these systems will become increasingly human-like.

Figure 7.4 shows how steeply the user base for such assistants will climb over the coming years.

Figure 7.4   Growth of virtual assistants

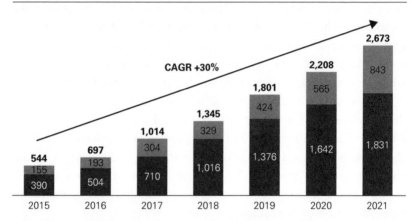

■ Number of Active **Consumer** Virtual Digital Assistant Users (in Million)
■ Number of Active **Enterprise** Virtual Digital Assistant Users (in Million)

SOURCE © Accenture based on Statista

The consumer world is showing the way forward for industrial and other product makers. In the industrial sphere, OEMs such as Tesla, BMW or Jaguar Land Rover, and tier-one suppliers such as Faurecia have built Amazon's Alexa and other intelligent assistants into their car interiors. If other traditional product makers can leverage AI in a similar way, they will dramatically boost their top-line growth. The idea is to create smart connected products and complementary services that continually enhance customer experiences. In the 'Products in Action' section of this book you will find many more examples, analyses and learning outcomes around this important topic.

One example of a new customer experience is a domestic AI assistant that develops recipes based on ingredients available in the kitchen, the pantry or the fridge at any given moment. This assistant also automatically orders and pays for other pantry staples when they're out of stock and updates the family on the status of dinner preparations. Such technology has the potential to change meal habits and times forever and there is no reason why AI couldn't transform other aspects of people's lives in a similar way – at home, in the car, at work.

Industrial businesses should, however, also see all these developments in the consumer sector as a warning. Over time, AI will cease to be perceived simply as a technological tool. Just as Alexa is now becoming the face of Amazon, there is a risk these AI brands could come to eclipse the brand identities of product makers who seek to exploit their technology.

AI in every product

Voice control via digital AI-enabled assistants is just the start of an unstoppable trend to smarter and more autonomous devices. In addition to the improvement of technology in the voice area, there are equally exciting AI advances in video, image and sound. For example, a Nest video doorbell can now use facial recognition software to identify who is at the door and then unlock it or not. And in the emerging market of autonomous cars, manufacturers combine a wide variety of AI camera and ultrasound technologies to design self-driving autopilots. These use cases indicate how AI is becoming the heart of digital services around smart products.

The trends towards more product intelligence will very likely go down the route of semiconductors. Moore's Law famously predicts this device category progressing as follows: ever-increasing processing power via ever-cheaper chips. As we have said elsewhere in this book, the processing power of an average connected device today surpasses that of a supercomputer 20 years ago by several orders of magnitude. The newest chips from manufacturers like Nvidia, Intel and Qualcomm are essentially embedded AI supercomputers for IoT products that deliver an astounding one teraflop of processing power on a module the size of a credit card.

Revisiting our Intelligence Quotient (IQ) framework, every product manager in the world should ask this critical question: 'How smart should my product eventually become?' The answer is, 'it depends', but the general trend towards increased intelligence is very clear. In some cases, a basic connected product IQ will be sufficient. But we believe most products will move to what we refer to as an Intelligent Product with significant AI technologies and a higher IQ rating. A smaller but increasing subset will move to a fully autonomous mode where the product adopts the role of a self-sufficient decision maker – for instance, when an implanted medical device measures blood sugar levels, does its own calculations and then administers the right dose of insulin.[18] 'Most devices will become more intelligent. But we will also see both fairly basic, low-cost devices whose intelligence is located in the cloud, and also 'hub and spoke' models with smart, inexpensive devices linked to one more powerful device nearby', says Marco Argenti, Vice President Technology, Amazon Web Services.[19]

It is clear that the latter example – self-commanding robotic technology, working under its own independent 'mind' and taking autonomous decisions – offers bigger value potential and market exclusivity for its makers and users, compared to less cognitively developed products operating in the same space. The difference is between mere service – letting you know when an insulin dose is required – and the fully realized outcome of steadily maintained health.

It is fair to say that modern AI technologies amount to a quantum leap not seen since the days of the computer hardware revolutions in the second half of the last century. It will likely be the breakthrough technology of the 21st century and provide a key way of getting product-making sectors of all kinds back onto a lasting expansion path. It is important for product makers to understand this and decide which path they want to be on.

Given the affordability of the various technology mixes enabling AI, it is only a matter of time before it becomes ubiquitous. In the end all industrial products will be reinvented with AI and move as much as possible towards autonomy, from medical equipment to pumps to industrial machinery to construction equipment to cars. The newly won intelligence will allow such products to communicate, adapt to context and exchange data via software platforms across the entire industrial value chain. The result will be a complete digital reinvention of the B2B and B2C product ranges we know today.

Those who get there first in their categories will achieve a lucrative head start as their products tap new value-rich digital markets. Manufacturers of any hardware should therefore look to prevent challengers such as start-ups and software companies from eating their lunch.

The AI pathfinders

Our research shows that a clear majority, close to 70 per cent, of manufacturers see AI as a key enabler of their product innovation and growth agenda in the coming years. In fact, 73 per cent see AI as unavoidably penetrating or transforming all industrial products and services. That figure rises to 91 per cent in China and 96 per cent in the United States, while it remains a surprisingly low 51 per cent in Germany. More than half state that at least 30 per cent of their product and service portfolio will be AI-enabled within the next three years (by 2021) and 25 per cent aim to have half of it AI-enabled by then.[20]

It is very important to note that almost all respondents – 98 per cent – have already begun integrating AI into their products in one way or another, though mostly in combination with other technologies, not as a singular product component. And, as the chart below indicates, manufacturers clearly understand that by combining AI with other digital technologies – especially mobile computing and big data analytics – they can drive both higher operational efficiencies and more differentiated customer experiences.

So, belief is not the issue – yet many industrial players still seem to struggle to realize their AI dreams. Although most know they need to change, only 24 per cent recognize that digital reinvention drives their top- as well as bottom-line growth. More than 75 per cent still take an unstructured, scattergun approach to the task, throwing resources from all parts of the organization at it. Most, moreover, are leveraging only

Figure 7.5   Combination of digital technologies with AI to drive differentiated customer experiences

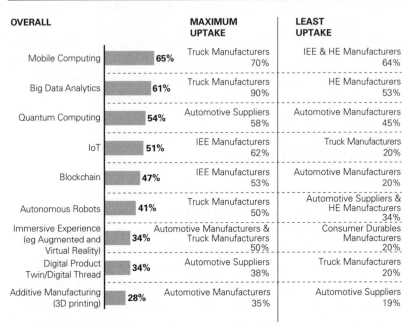

OVERALL		MAXIMUM UPTAKE	LEAST UPTAKE
Mobile Computing	65%	Truck Manufacturers 70%	IEE & HE Manufacturers 64%
Big Data Analytics	61%	Truck Manufacturers 90%	HE Manufacturers 53%
Quantum Computing	54%	Automotive Suppliers 58%	Automotive Manufacturers 45%
IoT	51%	IEE Manufacturers 62%	Truck Manufacturers 20%
Blockchain	47%	IEE Manufacturers 53%	Automotive Manufacturers 20%
Autonomous Robots	41%	Truck Manufacturers 50%	Automotive Suppliers & HE Manufacturers 34%
Immersive Experience (eg Augmented and Virtual Reality)	34%	Automotive Manufacturers & Truck Manufacturers 50%	Consumer Durables Manufacturers 20%
Digital Product Twin/Digital Thread	34%	Automotive Suppliers 38%	Truck Manufacturers 20%
Additive Manufacturing (3D printing)	28%	Automotive Manufacturers 35%	Automotive Suppliers 19%

IEE = Industrial & Electrical Equipment
HE = Heavy Equipment

their immediate ecosystem, rather than an expanded network of external start-ups, suppliers, customers and academic institutions that can complement and strengthen their existing capabilities. As a result, data quality and cyber security loom large among their challenges, as can be seen in Figure 7.6.[21]

And when it comes to transforming conviction into a commercially viable vision, only 16 per cent of survey respondents qualify, as detailed in Figure 7.8. In many cases, it's the CEO who puts the data-driven stake in the ground, authorizing top teams to start developing the investment and ecosystem strategies to acquire, process and secure the data needed to drive maximum value from AI. These industrial visionaries can see the big picture: 82 per cent rank enhanced customer loyalty and deeper

Figure 7.6 Major challenges for industrial manufacturers when embedding AI and digital technology in their products and services

Data Quality	51%
Data/Cyber Security	45%
The Choice of Making Versus Partnering Towards Developing an AI-Embedded Product/Service	45%
Data Sharing/IP Protection	40%
Customer Receptiveness	39%
Defining AI Algorithm/Code	37%
Human-Machine Interaction Complexity	36%
Regulations	35%
Job Loss Fear among Employees	33%
Others	11%

■ Foundational Challenges

insights from product and service usage as key value drivers. And the same proportion say that both greater safety and smarter solutions and services will be critical outcomes for their customers.

So, no product maker is already a perfect architect and practitioner of an AI-driven market strategy that makes the most of smart connected products, but a few businesses are already quite advanced on their journey.

As we have already noted, right now, products and services that use AI to integrate products seamlessly into users' lives – and delight them with truly personalized experiences – are most evident in consumer sectors. The big tech disruptors of Silicon Valley have set the standard here, but there are impressive vanguard examples to be found in the industrial sectors too.

Take 3M, the US cross-industry behemoth manufacturing more than 55,000 products for a broad spectrum of business and consumer users. The company's 360 Encompass System deploys computer-assisted coding and clinical documentation for the healthcare industry and is

Figure 7.7   **From believe to execution: a long way to go**

| Believe | Believe & Envision | Believe, Envision & Commit | Believe, Envision, Commit & Execute |

powered by natural language processing. This tool automates the process of extracting clinical information and concepts from unstructured data, such as the freely formulated text in electronic health and clinician reports. This information, which makes up 80 per cent of a healthcare organization's data, often goes untapped. Hospitals utilizing 360 Encompass have seen their net revenues increase by 19 per cent on average.[22]

Or consider industrial and mining equipment maker Caterpillar. A steep rise in its market value followed the deployment of CatConnect, its smart-equipment-based productivity service built on the back of AI technology, providing information and insight on heavy equipment use. Among other parameters, it monitors location, fuel consumption and usage patterns, but also offers tyre monitoring for maintenance management, and off-board safety reporting.

Another example is the German engineering group Bosch. Through its start-up, Deepfield Robotics, Bosch has debuted an AI-driven autonomous farming robot – the Bonirob. It can navigate down to a centimetre through crop fields using satellite navigation and uses machine learning to teach the robot the shape of invasive plant species leaves for removal.[23]

Similarly, specialist engineering firm Blue River Technology's robots combine computer vision and machine learning with their robotic system, LettuceBot, to apply fertilizer plant by plant where needed, and to eradicate weeds.[24]

AI has simply limitless potential for deployment cases. Take water usage in commercial buildings, where plumbing and failure of water systems remain largely a black box. There is still little data available even on basics such as water pressure, temperature or flow – even in modern plumbing systems. Leakages are therefore extremely difficult to locate but are estimated to represent a significant percentage of water consumption. So, US-based plumbing product manufacturer Symmons, featured in a case study in Chapter 12 of this book, is about to develop new business and product lines around smart and connected plumbing components.

Or consider finally the well-known global consumer goods company that decided to transform its coffee vending business in India by making all its existing enterprise coffee machines smart and connected. AI enabled them to enhance customer experience, personalization, uptime and brand recognition. They retrofitted all their 13,000 machines across India at marginal cost, driving their vending business to the next level by reinventing their product.

Product-making companies that successfully execute on their AI commitment display a few unique attributes: they build business models with a lifetime perspective; they aim to drive both below- and above-the-line value for their customers; and they do so by pivoting to a mix of AI technologies best suited to customer goals, and by building the necessary digital skills and underlying ecosystem.

These companies know that ultra-smart products and services are the future of industry. The time to join them on the journey to digital reinvention is now.

Time to get artificially smart

As we have hopefully convinced you, a tsunami of smarter, more connected and more AI-enabled products is coming. This force is only going to accelerate and get stronger over the next 5 to 10 years. We have seen that there is very strong user demand and acceptance of these technologies, so product makers have no choice but to redesign their products to incorporate AI. Increased AI usage will eventually lead to increased analysing and validating of AI behaviour. In the field of autonomous cars,

Figure 7.8   Visionaries focus on customer relevance and customer value creation

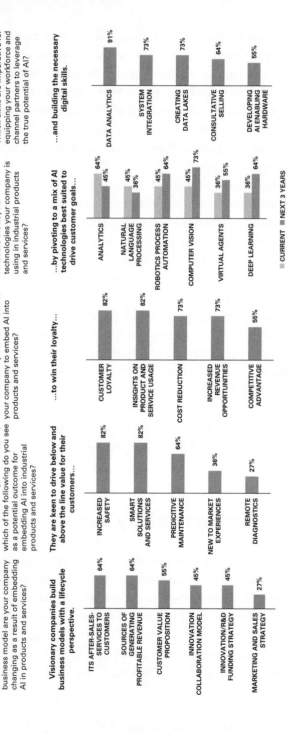

observing how autopilots behave can instruct engineers about new hardware or software functionalities.

But who will benefit most from the AI revolution? Is it traditional product makers or the suppliers of the AI platforms and the underlying semiconductor technologies? The market leaders in AI chipsets are Nvidia, which in July of 2018 had a market value of $150 billion, and Intel, with a market cap of $240 billion. The leading makers of voice assistants are Amazon and Google, which both have a market cap of about $900 billion. This stands in stark contrast to many of the more traditional product companies, the market valuations of which have stagnated. It is time for them to catch up.[25]

Takeaways

1 AI technologies will rapidly be adopted in the majority of products, making them increasingly intelligent with abilities to sense, comprehend, act and learn.

2 Most product companies are still at the very early stages of incorporating AI technologies into their products. While close to 70 per cent of industrial companies believe that AI will transform their products and services, only 16 per cent have to date articulated a clear vision for AI and even fewer have a committed and funded roadmap.

3 Product companies need an increased sense of urgency to build AI capabilities and embed them in their product and experience roadmaps.

8

Big Shift Five: From linear to agile engineering in the New

CHAPTER SUMMARY

....

The advent of smart connected products implies big changes for internal processes in business organizations, particularly the product innovation and engineering function. This chapter starts by describing the new R&D capabilities that will be required, goes on to explain the concept of new engineering, discusses the implications for marketing and brand management, and introduces the ramifications for supply chains and ecosystem management. Lastly, we discuss the concept of a digital thread and its criticality to the future.

....

The need to create smart connected products that deliver the strongest, most responsive possible user experience has major ramifications for how product manufacturers are organized. The way products are ideated, conceptualized, manufactured and eventually managed by makers and users must alter drastically.

The scale of the change cannot be overstated. Virtually every business process, organizational structure and culture code will be progressively and entirely transformed.

As a fundamental feature of the future, product-generated data about usage patterns, physical conditions and user demands will permeate organizations around the clock as the biggest source of value. Business organizations will only be able to handle this effectively by giving up on strictly linear internal structures, processes and hierarchies along their value chains. Instead, businesses will need information and data loops connecting different functions, as well as much greater agility and ever-shorter response times. These will be the only ways for businesses to extract significant insights, and hence value, from the huge data inflows at pace.

Figure 8.1 Engineering approach for the product of tomorrow

Product of Yesterday:	Product of Tomorrow:
R&D – predictable and incremental	R&D – unproven technologies and solutions, shift in value sources (from product to services to outcomes) ... increase in R&D budgets (2- or 3-fold for OEMs)
Innovation – incremental, evolutionary in terms of technology, capabilities	Innovation – transformative (disciplines, approaches, budgets....) especially renew capabilities as digital technologies permeating the product
Approach – linear, measured in years	Approach – agile, sprinted with multiple clock speeds – R&D for the core vs the New, R&D for hardware vs software vs experience
Contained in the enterprise	Open to the outside – ecosystem for innovation and access to new capabilities

The path from linearity to 'experiment and scale'

Let's quickly remind ourselves where most product companies are coming from. Traditionally, development required mechanical engineers to create physical variants at significant financial cost. Innovations would often be based on user insights, usually derived from focus groups or watching users in the field. The resulting new features would have a shelf life of several years. New devices would be put into production with the handover of a physical product prototype plus ready-to-manufacture blueprints to manufacturing colleagues who then organized assembly lines for production.

Now, to see where product companies are going, let's start with innovation and engineering, which for smart connected products is a very different ballgame. This is one of the most important changes that companies must take on board as it has implications for the whole manufacturing process and practically all other business functions, and more broadly the whole enterprise.

In many manufacturing sectors, software is already eclipsing hardware. This immediately injects speed and agility into development, enabling the creation of device variants instantly and at very low cost. Product profiles can be made to change, or can even change autonomously, by recoding their software anytime, often remotely. But if not managed appropriately, this added software can actually increase costs if it results in a wider array of product variants with different software versions or derivatives in the field at any given time. This explodes the need for installed base testing and compatibility checks.

For instance, John Deere, the agricultural machine maker, now produces engines in which it's the software not the hardware that delivers the horsepower, allowing engines of different strengths to be much more quickly and easily produced and shipped.[1] For domestic appliances such as washing machines and dishwashers, users now commonly download software that, say, adds more programs or adapts machines to local water qualities.

The constant data loop fed by the installed product base gives development engineers a treasure trove of insights. Many of these, as we have seen, can be leveraged for existing products via software updates, delivering an instantly improved user experience.

But the data loop also opens up room for freer experimentation with no immediate goal of product improvement. Engineers might come up with completely new product categories that could not have existed even five years ago, such as Amazon's AI-enabled smart Echo home assistant. Or they might create new digital services complementing the product, such as audio equipment maker Bose's 'find my buds' function.

To make the most of the opportunities, processes and teams must become as agile as possible. Bringing this about requires a major cultural change and significant enhancements in capabilities that many companies struggle with. Figure 8.2 illustrates the magnitude of this transition. It contrasts old-style linear approaches with so-called Exponential Organizations (ExOs), businesses which have systematically implemented accelerating technologies for their advancement.

Figure 8.2 Characteristics of linear and exponential organizations (ExOs)

Linear Organization Characteristics	ExO Characteristics
Top-down and hierarchical in its organization	Autonomy, social technologies
Driven by financial outcomes	Massive transformative purpose, dashboards
Linear, sequential thinking	Experimentation, autonomy, agile is the norm
Innovation primarily from within	Community & crowd, staff on demand, leveraged assets, interfaces (innovation at the edges)
Strategic planning largely an extrapolation from the past	Massive transformative purpose, experimentation
Risk intolerance	Experimentation
Process inflexibility	Autonomy, experimentation, agile is the norm
Large number of FTEs	Algorithms, community & crowd, staff on demand
Controls/owns its own assets	Leveraged assets
Strongly invested in status quo	Massive transformative purpose, dashboards, experimentation

SOURCE: © Accenture based on ExO Works

Alongside more conventional research and development, there must be a culture of 'experiment and scale'. A new form of development circularity is needed. Ideation processes and ways to conceptualize and adapt smart products in fast-paced sequenced editions will be the norm and innovation will be underpinned by looped trial-and-error sprints as well as a pioneering 'fail forward' culture. Research and development teams will have to embed new immature digital technologies in their products which often means overcoming a mental hurdle as new technologies bring new uncertainties to the process and into the product.

A common barrier here is that new smart connected products require a much wider array of skill sets than conventional products. In addition to mechanical and industrial design engineers, there is for example a greater need for design thinking experts, software engineers, experience designers and experience engineers, IoT and connectivity engineers, cyber-security experts, AI specialists and so forth. In many businesses these pointed qualifications are not sufficiently available within the organization or might be scattered across various functions, not aware of each other. Often these skills do not exist at all in organizations.

Moreover, the type of cross-functional collaboration required is rare today. Commonly, different disciplines are based in different buildings or even cities. Some functions may even be sourced from external service providers, with their attention divided between various projects. All this can slow down the ever more important time to market in the digital era.

Therefore, businesses that change their innovation culture along the lines described can point at massive time wins. Schneider Electric, the electronics equipment maker, cut the time to market from two years to only a few months. And more agility through using 3D printing lets carmaker Ford create a prototype within four days, where previously engineers had to wait four months.

Steve Myers, the CEO of Mindtribe, a hardware engineering company working with many leading companies in Silicon Valley, is very candid in his assessment of what he sees in companies that have not yet started to inject more agility: 'Connected hardware development is broken today. Way too many hardware development efforts result in products that aren't very good, never ship, or take too long to develop (which also costs too much).' He believes an agile approach leveraging new technologies is required:

> 'The core goal of agile process is to get experiential prototypes in front of people quickly, as well as to validate everything else that's most important about a product as early as possible. With the advances in technologies like 3D printing we can get new prototypes out every day, sometimes multiple iterations in only one day. In the future, we will likely be able to apply virtual or artificial reality technology to enable this experiential prototype for people without actually making a physical product.'[2]

Ideate, launch and iterate in cycles

Three core pieces of contextual product information are now available around the clock: where, how, and by whom products are being used. Some companies use this just to better meet customer needs and improve revenue. Others do more, connecting the information with further shared data they own or access through ecosystems. This combinatorial intelligence allows them to design the next level of products, with even more enhanced features and services.

The Internet industry should be a source of inspiration for incumbent product engineering firms. Cloud platform engineering at companies like Amazon and Google is agile, non-linear and fast. Product innovation at these companies is grounded almost entirely on incessant digital experimentation with 'living' services such as social media, search or video portals. The goal here is to launch a so-called 'minimum viable product' (MVP) as fast as possible and then learn as much as possible from its early life in the market. On that basis successive rapid-fire updates are developed, usually resulting in new functionality as often as every week and sometimes even every day.

Smart connected products will increasingly be the by-products of real-world testing and experimentation, and businesses will have to develop structures to accommodate this new mode of innovation. Trying things out is as important as putting the focus on quality, and having many good ideas often turns out to be more valuable than having one great idea. 'If you double the number of experiments you do per year, you're going to double your inventiveness', says Amazon founder and CEO Jeff Bezos.[3] Software can always be effectively improved by real-world iterations in end-user markets. And the more product hardware is driven by digital fluidity, the more makers of physical products can also work according to this model.

There are almost legendary examples of how experimentation has led to huge profits. Amazon today generates around 35 per cent of sales on its platform via its recommendation engine, suggesting products to users based on their past shopping behaviour.[4] This is software Amazon stumbled upon in a freed-up innovation process rather than getting it from strategic development planning. Fast, cheap, iterative experiments and quick scaling stood at this innovation's cradle.

Flattened hierarchies and fluid organizations for greater agility

As is probably becoming clear, agile processes are incompatible with the waterfall hierarchies typical of traditional development environments and the differences are enormous. A traditional project manager coordinates a range of tasks across various teams. Under an agile regime the number of tasks is minimized, and with this the need for vertical coordination. The remaining tasks are handled horizontally among the agile teams themselves. So widespread delegation of decision and approval processes is necessary to inject speed into organizations – as is digitizing and automating, as far as possible, validation and testing of products.

Self-organizing, product-focused teams sit at the core of development activities around smart connected products. Only they can muster the agility for quick regrouping that is needed when products are so closely connected to their makers. In a more conventional set-up, software development teams would work under technology managers to translate requirements drawn up by software architects into technical specifications. Agile approaches can do without this needless translation step. Developers can even be empowered to talk directly to architects or even product owners in order to get a better understanding of customer needs and shape their software accordingly.

Team size can also turn into an agility killer. Smaller teams have a surer footing when pushing new ideas forward compared to bigger groups that are often held back by inertia and lack decision-making speed. Amazon's famous 'two pizza' rule is an example of this thinking. It states that no team should be larger than can be fed by two pizzas. Even if a project scales, the ensuing larger teams are broken down again into smaller ones in accordance with the rule.[5]

Product developers and innovators in a flattened organization should be given relative independence to pursue their explorations. Their experimental efforts will be far less successful if channelled only into pursuing management's ideas. Ideas from management are fine as optional triggers for experiments, offered collaboratively, but if senior personnel try to predetermine the outcome by prescribing specific goals, they will stifle the work, not strengthen it. The whole point of experimental work is that it be free to shoot darts in multiple directions, because no one knows for sure where the bullseyes are.

At Internet platform companies such as Google and Amazon, this agile innovation with a flattened hierarchy climate is firmly in place. Development personnel do not need approval to pursue their ideas, or to decide when to drop them. Traditional hardware makers, which often know little of such liberal but hugely value-driving processes, need to learn and follow them sooner rather than later.[6]

Chinese domestic appliances maker Haier has taken the principle of flattened hierarchy to the extreme by compartmentalizing a giant organization into hundreds of small entities with their own CEOs, profit and loss accounts and marketing responsibility. We cover this interesting move in an in-depth case study in Chapter 12.

'Engineering in the New' to achieve 10X digital factor

Building the new capabilities and implementing the approaches outlined in the chapter is a major transformation for R&D that we call 'Engineering in the New'. Collectively we believe these changes can result in a 10X or order-of-magnitude improvement in product innovation efficiency and effectiveness. Figure 8.3 summarizes these changes.

Achieving the 10X 'Engineering in the New' factor requires rethinking each step of the process. Start with the ideation process, which needs to rely more on crowd-sourcing and open innovation and co-innovation techniques with a larger, more open ecosystem of partners. The product and system architectures need to be fully redesigned in a more modular model to enable a platform engineering approach and easier integration of components from any source. The prototyping and development process must be agile and led by design thinking. The product design needs to take into account the lifecycle usage models and data required, particularly for as-a-service models. In short, this is a major transformation of engineering culture and processes.

This new approach requires new skills and new methods of collaboration. Mechanical engineers increasingly need to work with other disciplines such as microelectronic or software engineers, data scientists, service designers, platform and ecosystem architects and managers, digital stack and human-to-machine interface experts, and other niche competencies.

Traditional physical product companies have seen their need for software engineering teams and these other digital experts swell from a small percentage to often half their product development team. Car supplier Faurecia, for example, plans to double its quota of software experts in only a few years as it heads for the launch of its new Smart Life on Board line.[7] Many other industrial businesses such as GE, Airbus, Daimler or NIO are also hiring software engineers, system integrators and data scientists in droves, often putting them to work in design centres in Silicon Valley, where such expertise is the most concentrated in the world.[8]

Figure 8.3 The 10X digital factor in product development

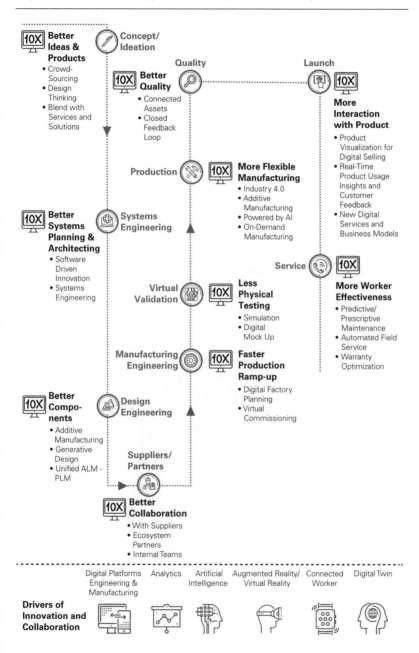

Several leading technology companies are extending this idea beyond design centres to create integrated, next-generation product innovation and production hubs that bring all the necessary skills under one roof to link them up using agile development methodologies.

The following chart shows a real-life business example in more detail. Electrical equipment group Schneider have lined up their product teams around three headline notions: desirability, feasibility and visibility. It is very instructive to see which micro-competency and task is assigned to which team to make the most of different specializations for an overall smart product development strategy.

The receding dominance of mechanical engineering is mirrored by the rise of digital user interfaces – for example to control a vehicle or an industrial machine. These interfaces are easier to modify than physical

Figure 8.4　**Multidisciplinary teams for the New**

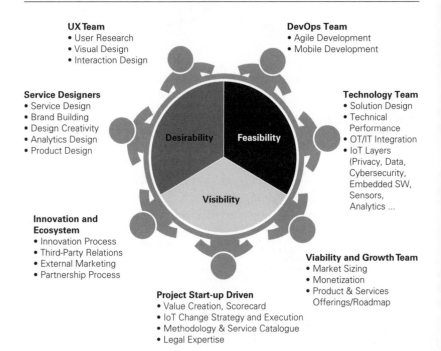

controls and they enable greater operator flexibility, allowing not only for much greater user convenience but, more importantly, much finer customer segmentation. Display set-ups become quickly alterable, eg to suit client groups with special operational requirements.

The Chinese appliance giant Haier, for example, has washing machines that can stand outside the house – as is customary in some parts of China – after being reprogrammed by clients via the device's touchscreen. Also, via the touchscreen, the appliances can be set for specific regional water qualities. Some devices can now be remotely controlled via a smartphone or tablet app, and some have even started involving augmented reality-based interfaces in which smartphones or smart glasses, pointed at the product, generate overlay representations with monitoring, performance or service information.

Matching the right spectrum of features for such a versatile spectrum of individual customer needs by combining software and hardware functionalities is a crucial new design and engineering challenge. To meet it, manufacturers of smart connected products must hire very focused specialists.

From ageing products to evergreen design

Everybody working in the development of these items needs to take on board the concept of 'evergreen design'. This refers to the change from creating discreet generational model versions towards continuous editing, continuing after purchase and throughout the product's lifecycle. Traditional approaches such as product launches and product lifecycle management go out of the window. Instead, design and engineering teams working in development units need to be clear about their lifelong supervisory role over smart connected products.

The big pioneering Internet leaders, companies such as Amazon or Google, already employ platform managers who own every aspect of a product and the product experience over its entire lifecycle. This is a dramatically different role from that of traditional product managers who focus on getting to a successful product launch and then move on to new product development as soon as possible. In a hardware product company, this platform engineer role is even more critical as they also need to interact with both manufacturing and the installed base field Services.

With manufacturing, changes to an evergreen product's hardware or software need to translate into an immediate reconfiguration of production. With the services departments, field engineers need to be updated and trained on a continuous basis. This holistic thinking approach is referred to as 'Design for X' or 'Develop for X'(DevX) with X standing for a variety of areas such as manufacturing or support. Particularly once services are offered on the back of a product, someone must take up the overall responsibility for all aspects of the offer's lifecycle, otherwise it can be in danger of coming across as disconnected and not obtain the level of customer experience required. This product management role should also be responsible for increasing the value over time for the consumer.

It is worth describing this example of the iterative process in more detail. In developing its autonomous driving functionality, Tesla uses AI software to machine learn the best of human driving skills. The still manually driven cars are pitted against software simulations running in the background on the car's computer. Only once the background program consistently simulates moves more safely than the human driver is the autonomous system considered ready for its real-life debut. At that point, Tesla releases the program to the fleet through a remote software update.

Note what a revolutionary change this is. Traditionally, products have been tested using guesses about how customers would use them in the field. Now, continuous monitoring of real-world performance and usage data allows product makers to see and fix design problems that testing missed. Looking at Tesla again, the carmaker realized that some of its batteries could catch fire after being punctured by objects whipped up from the road surface. A single remote software update raised the suspension sufficiently to avoid such incidents.

Yet releasing smart products that are well tried and tested can be vital to deliver a functional experience, free of irritations. And digital technology, with tools such as 3D representations delivering key insights, allows for such comprehensive testing, prototyping and simulation prior to market launch, compared to the old product world.

Methodologies such as Scrum are key to achieving agile success. The concept has spread like wildfire across the automotive industry for both hardware and software development. The aim is to keep a product in a constant state of redesign depending on market feedback. Its early pioneers were inspired by 'inspect and adapt' approaches to coping with risk. Information from the real world informs decisions in scrum processes, and products get a regular design overhaul in short sprints of one or two weeks.

As-a-service business models determine hardware features

Agility and acceleration are especially important when smart connected products start to deliver as-a-service business models that take over from the transactional selling of discreet products. This will often require significant changes to sales, fulfilment, provisioning, governance and support systems in order to match requirements for security, scalability, operability and compliance.

Devices used to deliver paid-for services must be enabled to record usage data for appropriate usage billing. Design engineers must therefore have a clear idea what the billing criteria are going to be and attach sensors accordingly. It makes a huge difference, for instance, if a printer bills a user per printed page or per millilitre of ink, with differing hardware set-ups required in each case.

In some cases, development engineers must also bear in mind that once a smart product is delivered as a service the cost of maintenance stays with the product maker, and this changes design parameters substantially. That's why bike-hire schemes, for example, use especially robust technologies such as chainless drive shafts, puncture-resistant tyres, product durability.[9]

Agility becomes even more paramount when smart connected products form platforms surrounded by ecosystems. Partners in the ecosystem must have seamless access to the data loop the product is embedded in to deliver their piece of the co-created user experience.

System interoperability at software and hardware levels therefore becomes eminently important for products. Developers must think carefully about how to craft that access. When the product delivers an outcome that sums the contributions of ecosystem partners, product design may need to be a matter of collaboration with those partners. Development teams have to be mentally open and prepared for this kind of innovation mode – and there may well be teething problems, given what they're used to. To provide a sense of the kind of partnership they may have to accommodate, domestic technology specialist Nest's smart thermostat can communicate data with smart door locks made by Kevo, another home technology specialist, which is a fierce rival in other markets. In Amazon's ecosystem for ebooks, on the other hand, the online retailer works with the maker of its Kindle tablets and an array of book publishers, but keeps the partners separated.

In many settings today, needs are in flux and aims undefined. Potential partners require orchestrators who can find mutual benefits for all participants and encourage them to work together to identify new business opportunities. The job is to create ecosystems that are adaptive, where partners develop projects or innovations together in agile configurations, rather than centralized ones where a single partner dominates.

Integrating engineering and IT for agility

What must not be underestimated is the integration required to achieve agility and speed. A data-reporting product must be connected to the internal and external information technology systems of its makers at many points. Smart connected product capabilities – such as looping data, configurability, real-time monitoring and liquid feature upgrades – require heightened levels of support, security, operability and performance from back-office enterprise systems. This is not easy to achieve when a business comes from the old product world, says James E Heppelmann, CEO of PTC:

'The engineering teams don't know that much about the data concerns, the security concerns, the failover concerns and the cloud analytics capability of the business. That is a whole new world to teams who work on physical products. The IT people within the organization know a lot about that, but they never have been involved in a product delivery process.'[10]

To make the development of hardware components as agile as that of software components, businesses often use a separate Application Lifecycle Management (ALM) approach in combination with so-called Development Operations (DevOps) to structure and channel software development processes and make them as fast as possible. This approach includes a carefully constructed and integrated set of software engineering tools for everything from requirements and configuration management through to automated testing and deployment. This organization-wide method allows engineers to focus on innovation and avoids cost added to products when teams maintain their own insular tools. Software-driven connected devices also benefit when ALM tools used for software development are integrated with Product Lifecycle Management (PLM) tools that manage the actual smart product from concept to life in the field to retirement. Using the same toolset gives product development leadership the transparency and data they need to make informed decisions about supply, demand, portfolio and productivity; it also provides the insights needed to re-use components across multiple products.

Product development teams and chief information officers must collaborate to blueprint and implement these new requirements from the beginning as core components of internal IT architectures and services. The flow of data and transactions from new generations of smart connected products requires upgrades in infrastructure, software monitoring and operational processes. In many cases, the size of the required upgrades is massive and orders of magnitude beyond historic needs.

Legacy IT systems typically struggle to adapt to this new product world as they were never designed to deal with the magnitude of transactions, data volumes and processing power now required. One technique being deployed by many leading companies is to separate IT systems into two components. On one side are so-called 'systems of engagement' that interact with the customer or user. These must be digital-based systems with a high degree of flexibility and customizability. Product development and business teams work closely with these 'systems of engagement' to drive experimentation and innovation. The other component is 'systems of record' that hold the core data and transaction history, which can remain more monolithic and inflexible. The approaches to developing, managing and maintaining these two components are completely different.

All this means that information technology budgets and resources must be aligned with product development roadmaps in ways that were never done historically. New, agile product development practices mean higher time-to-market pressure, which must be factored into resource budgeting, bringing funding into sync with product development and launch times. IT teams might even be required to use more iterative ways of developing and delivering their solutions by adopting lean and agile ways of working from their product development counterparts.

Integrating IT and engineering also creates a significant governance challenge as typically these organizations are quite separated today, in terms of both formal organizational structure and culture. But as internal IT systems become key to the product and experience design, IT needs to become embedded in the decision-making and innovation processes as well. Of course, the reverse is also true, meaning that product development leaders will need to have more decision-making power over core IT systems, particularly the systems of engagement.

Twinning and threading your product

With the emphasis on reinventing conventional products as smart and connected devices and services, a core priority for product companies is to deliver new customer experiences around their software-enlivened hardware throughout its lifespan. Agile developer and producer teams, manufacturing engineers, the field technicians but also the sales and marketing teams must have, right on their digital drawing boards, real-life, real-time representations of any usage situation in the field. Today,

Figure 8.5   **From PDM to digital thread and digital twin to support new products and business models**

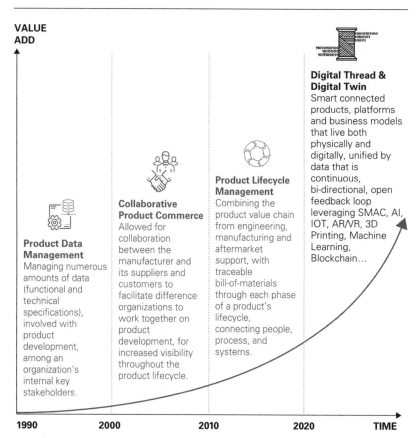

VALUE
ADD

Digital Thread & Digital Twin
Smart connected products, platforms and business models that live both physically and digitally, unified by data that is continuous, bi-directional, open feedback loop leveraging SMAC, AI, IOT, AR/VR, 3D Printing, Machine Learning, Blockchain...

Product Lifecycle Management
Combining the product value chain from engineering, manufacturing and aftermarket support, with traceable bill-of-materials through each phase of a product's lifecycle, connecting people, process, and systems.

Collaborative Product Commerce
Allowed for collaboration between the manufacturer and its suppliers and customers to facilitate difference organizations to work together on product development, for increased visibility throughout the product lifecycle.

Product Data Management
Managing numerous amounts of data (functional and technical specifications), involved with product development, among an organization's internal key stakeholders.

1990   2000   2010   2020   TIME

this is extremely challenging for most product companies as typically the data models and technology support systems are very siloed across engineering, manufacturing and service and support functions.

But this end-to-end mindset, supported by systems, is what is required to succeed in the future world of reinvented products.

Two related strategic concepts that address these issues are rapidly gaining traction in the world of smart connected products. One is the digital twin, which represents the product as a high-resolution digital facsimile as it currently is. The other is the digital thread, which tethers the product itself, its makers, and its users but also external ecosystem partners and sponsors throughout the product's life.

The two concepts derive from military practice, where they are used to deal with complex data flows and data structures stemming from the development and operation of combat aircraft. However, the concepts are also quickly gaining traction in industrial manufacturing. Nike shoes, for instance, can now be produced from a choice of 25,000 different features. Nike uses the twin to manage this kind of complexity with great success.[11]

SMART VOICES   Olivier Ribet, Executive Vice President, Dassault Systèmes

Let's think of a smart washing machine. A decision is taken to enter the Japanese market with the product. This machine has to be small, portable, easy to install in any city in Japan. The concept also says it has to be able to wash more than 200 times per year and it cannot cost more than 200 euros. It has also been decided that the device has to offer the functions of washing, drying and, as an additional feature, humidifying the room. Plus, this machine cannot weigh more than 150 kilos, has to withstand 200 Newtons of impact when it drops from a truck. All these features and functions – and in real development processes of course a lot more physical properties and logic functionalities have to be determined – can be represented in a 3D digital twin available to anyone having to deal with the product with modelling, simulating, producing, marketing or servicing the product.

With the digital twin you would simulate every aspect and episode of the real-life behaviour of the machine, down to little details. So you virtually press the start button, you virtually put water of a temperature of 75° Celsius and pH of 2.6 into the machine and then you watch what happens. Everyone involved can see digital continuity from the design to the mechatronics to the system to the software. All the disciplines are managed and governed together in one data model. So, it is not just some vague marketing dream. It is a perfect representation of all engineering aspects of the original that is not real yet, but can be. All components have been selected and you know what you will build as you have a perfect facsimile of what the people on the production lines will be assembling. At this stage you could, for example, bring in local electronic retailers and discuss with them whether they would put this machine with these defined specifications into their catalogue.

The digital is an 'always on', near-real-time electronic representation of a product or platform device, both hardware and software, showing its current configuration in each moment of development and use. This representation includes xCAD visualization and related engineering information such as product specifications, geometry models, material properties, validation results, IoT sensor readings and associated simulation information. The virtual twins use data harvested from sensors installed on smart product devices or product platforms to represent their status, working condition or position. These data files are analysed against business and other contextual data to uncover insights in a virtual environment, so that they can be applied to improve product experiences.

From that high grade of detail, practical conclusions can be drawn for mechanical and software development but also for future service propositions. As smart products become ever more connected, they – and their twins – become increasingly rich data platforms, offering growing sources of differentiated value. These platforms thus generate the potential for new, often third-party-provided, services, closer customer and partner relationships and, crucially, new revenue streams.

A digital twin for a smart connected product carries four fundamental benefits: a smart product's digital representation allows much more effective assessments of the hardware's current and future capabilities over its lifecycle. Product performance weaknesses can be spotted early by simulation results long before physical products and services are developed. The twin also heightens general operability, manufacturability, inspection, and sustainability. Design features and model blueprints can continuously be updated and refined in accordance with user preferences.

The twin approach has now been adopted by a wide range of product-making businesses, ranging from automotive manufacturers to makers of medical equipment, engines, pumps and washing machines down to 'simple' consumer products such as sports shoes. The advantages of this tool are mirrored in surveys we undertook. They show that most executives feel that digital twin first movers will achieve a 30 per cent new revenue upside, that 90 per cent are currently evaluating a digital twin concept for their existing or new products and services offering, and that the use of digital twin technology is anticipated to almost double over the next five years.[12]

The brother, if you will, of the digital twin is the digital thread. It is the umbilical cord of a smart connected product that is never cut once a device, coming from complex multi-sourced delivery, has been released into life in the market. The digital thread creates bi-directional linkages between the physical space and the digital space through processes and technologies. Such a thread extends the concept of the digital twin into a product's entire lifecycle to encompass ideation, design, engineering, performance, realization, manufacturability and serviceability as well as retirement. It enables the development and update of the product's digital twin, from concept to grave, leveraging analytics, IoT and other advanced technologies.

The thread also homogenizes a company's different IT systems with regard to the smart connected product. It connects all participating units, parties and digital contexts with which a product or service interacts. In that regard it allows a connected data flow and integrated views throughout a product's lifecycle and across traditionally siloed perspectives. Thus, it forms the overall basis for the effective iterative development processes characteristic of the digital age.[13]

It is worth mentioning in this context the rapidly growing weight of Circular Economy business approaches, a burning corporate topic as resources for the industrial sphere are dwindling and the need for reuse increases.

The Circular Economy does away with the 'make, use, dispose' approach to production and consumption, in favour of prolonging the use of natural resources, products and services as well as eliminating waste.

Figure 8.6 Digital across the product lifecycle

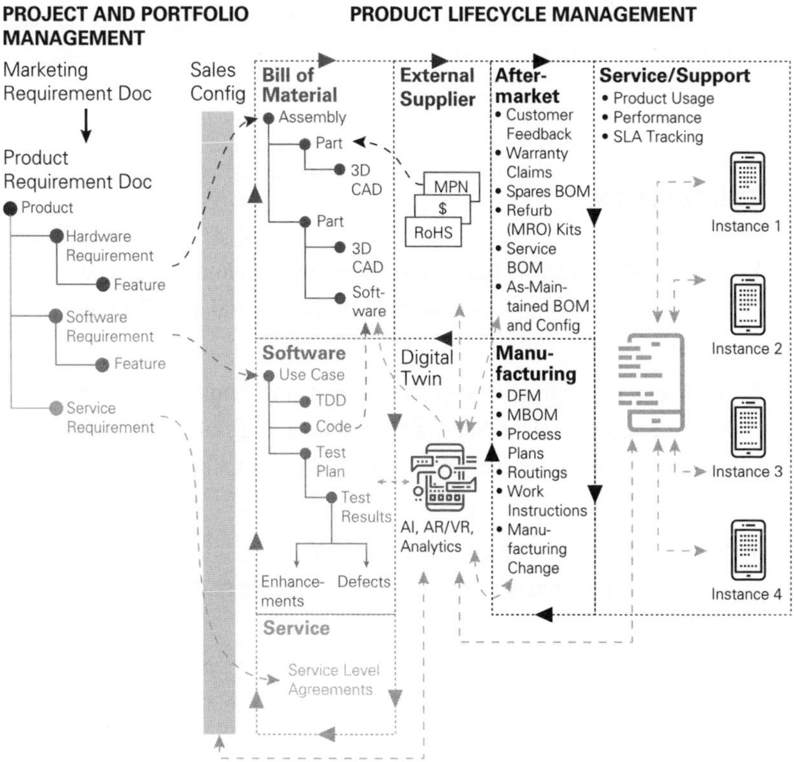

Thus smart connected products are themselves becoming powerful sustainability drivers.

For businesses, that means doing more with less. In a world of smart connected products this question turns critical as a lot of input resources can be spared when hardware products get an extended lifecycle through frequent software updates. More and more product-making companies are therefore integrating a Circular Economy mindset, which has implications on how smart connected products are designed, engineered and maintained after shipping.

Products must be principally designed so that they can offer value and services in a circular way. And their value and user experience must also be maximised not only by a single use but also by repeating uses until the end of a product's life.

Takeaways

1 Traditional hardware product development is broken and will not work in a smart connected world. In the new world it is all about agility, iterations and experience.

2 A complete transformation of innovation using 'Engineering in the New' concepts and methods is required. Successful implementation can yield a 10X improvement in product development efficiency and effectiveness.

3 Unified data models and digital threading across the enterprise are required to enable next-generation products and as-a-service models.

PART THREE

The journey to the reinvented product

9

Seven pivotal capabilities for managing the reinvention of the product

CHAPTER SUMMARY

....

Product-making companies can reinvent themselves through building the seven capabilities outlined in this chapter. Building these capabilities will help them unlock new value from their existing products and tap the new markets for smart connected products. Furthermore, these capabilities will enable companies not just to develop smart connected products one-off, but also to loop back the resulting usage-based findings into the development of the next and following generations.

....

The product of the future is smart and connected, with the ability to be responsive, adaptive and collaborative. These traits constitute the product's prospective value. Pure hardware-based products fall shorter in terms of value add for their customers and hence lose significance. Software-driven products with respective services generate the value propositions of the future based on digital intelligence as the principal driver of market success and differentiation. Take as an example the dashboard of a car. Historically, this was a classic electro-mechanical component, but dashboards are increasingly becoming a digital interface that is highly customizable and remotely updateable.

This transition to a fully reinvented product is characterized by – as we see it – the five big shifts, as laid out in the previous chapters of this book. To recap: these shifts are interlinked. First, a device's mere feature profile steps back in the perception of the user to make room for comprehensive product experiences and outcome delivery. Second, the hardware pales in economic value compared to value-rich services that are enabled by software and digital intelligence. Third, formerly insular products turn into platforms complemented by other technological components or services. Fourth, its behaviour and inner works, while in use, move from mechanical functions to control by artificial intelligence. And fifth, the way a smart connected product is produced changes from a linear value chain to looped iterations in agile development and manufacturing processes.

The necessity arises for new competences – in management, workforce, and the organization of business processes. That's why we present here seven capabilities we believe are the minimum for starting the journey towards smart connected products.

1 Design 'flexagility'

The first capability is 'flexagility' in designing the product of the future. Flexagility combines flexibility, the willingness and capability to change, and agility, the speed of change. The implication is of quick, non-linear development workflow with the ability to wipe the slate and think again if an idea, project, product or service does not seem up to the impeccable user experience it is meant to deliver.

Embark on an iterative product development approach

Flexagility strongly implies iterative design approaches. Out goes the old-style programmed linear development process with a firm sequence of steps for a product project – from ideation to market launch. The designing of a product rather needs to be moulded into a revolving and circularly revisiting but forward-looking design practice for which 'design thinking' methods are the most efficient frameworks.

With its roots in product design, design thinking is a perfect methodology for finding and developing services around products that customers really want. The guiding focus throughout is the end user, so this needs to become an integral part of design and development from day one, with things like interviews, observations in the field and techniques such as 'customer journeys' forming the basis for new offerings. In addition, development teams will be multidisciplinary so as to be able to collaboratively provide all the necessary inputs. A key prerequisite is therefore to break down the silos that usually separate relevant parties. Technical feasibility, economic viability and desirability for users can, in this way, all be factored in and developed in tandem.

The methodology, applied in the version we recommend, comprises four stages executed in a fast and agile manner: discovery; description; ideation, prototyping and testing; and finally implementation of a service or product design.

User testing, rather than coming only at or near the end of development, occurs iteratively throughout, allowing problems to be ironed out quickly and cheaply before arriving at the final product or service.

The chief designer of streaming platform Netflix, Andy Law, says, 'Depending on the complexity we may test an idea multiple times to identify what, if anything, is contributing to a positive member experience.' His mantra is 'Test, learn, repeat'.[1]

Figure 9.1   From discovery to implementation

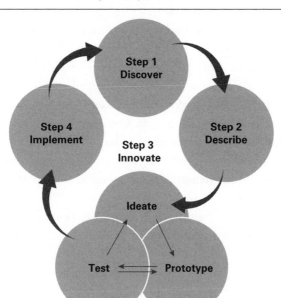

Much of this has been in use at leading design practices for some time, but it really comes into its own in the New. It means prototypes can be digital, but more importantly, it means that, via data feedback, user insights continue to inform developers even after sale. Crucially, because the product/service value is now digital, those insights can also improve the offering after sale, via software updates. That's why design thinking is the way forward when it comes to getting to a smart connected product that is evergreen throughout the product lifecycle.

Customer-centricity across organizations

Achieving user- or customer-centricity in smart connected product designs requires fundamentally different skills compared to traditional product design. You will need to bear this in mind when putting together multidisciplinary teams for design thinking. As we have indicated, in this new world, design will be more data-driven, leveraging the tremendous amount of feedback generated by the smart device. Design must also extend beyond traditional product usage thinking to more holistic service experience design once the latter has become the value driver of a product. Only if the stiff sequence of 'ideation, conceptional design and prototyping' is dynamically looped to become reactive to new needs and instant market feedback can you be customer-centric around the clock, 365 days a year. Consider as a good example the Italian industrial machinery maker Biesse, which has connected thousands of its specialized wood and stone processing machines working around the world for remote assistance and maintenance services provided to customers.[2]

The digital twin, a concept we explained in detail in Chapter 8, is a useful tool for this. As a permanently available digital representation of a smart device, even when it is in use after being shipped, it informs and helps everyone involved in a redesign iteration process visualize what works and what does not. Such advanced forms of product listening are the basis for value-creating services. Olivier Ribet from Dassault Systèmes, a software company that specializes in collaborative engineering solutions, says:

> 'The technology used to model and simulate an experience end-to-end
> makes you fast as a creator and manager of a product. It orchestrates every
> step and weeds out processes that you do not need. It also can cut out a
> lot of linearity. So, designers, engineers, factory workers, distributors and
> marketers all work with one single version of the truth.'[3]

The simple rule of thumb is: be data-driven to become customer-centric. Andy Law of Netflix again: 'Research and data drive a lot of what we do, but even if we don't start a project based on a data point, our data science and consumer insights partners are always involved in our explorations.'[4] It's hard to track improvement if you don't start with a clear benchmark. If you don't have the data internally, leverage your ecosystem to fill the data gap. Know what you're measuring and where you're starting before you ship.

But use inputs other than just pure data, like qualitative impressions on the user experience, to confirm that you are going in the right direction.

It is not for nothing that big data-driven organizations such as Amazon or Google build experience centres where customers can have hands-on interaction with products to provide insights that complement mere data findings. Channel information, taken from distributors such as retailers, can also enrich your customer insights, as can ethnographic research, and crowd-sourcing ideas from users.

Equally important in the context is social listening, the collating of user discussions about a product. This requires identifying diverse online communities, forums, hashtags and blogs dealing with the topic and screening them with search tools for issue-based discussions and consumer statements on product experience achievements. Social listening is particularly helpful at identifying the 'moments that matter' in the customer experience.

Take as an example the agricultural website https://community. agriculture.com. In its talk section, farmers and other equipment users discuss their experiences with a wide variety of equipment. There's a very large 'machinery' subsection, and also sections on 'computers and more' and 'precision agriculture'.[5]

Increasing flexibility and agility in product design becomes especially important as consumers are influenced by social media and hence can unexpectedly shift their needs. Professional communities, interest-based crowds as well as expert forums increasingly impact the end user. The willingness to adjust the product design accordingly and the speed to act upon it are crucial attributes for designing the product of the future.

Open innovation in the ecosystem

In the digital New, multidisciplinary teamwork on design and development will increasingly involve not just individuals from within one organization, but also from ecosystem partners. Developing, enabling and nurturing an ecosystem with the right partners becomes a critical success factor. With the necessary internal processes and approaches in place, companies will have a base from which to effectively collaborate with universities, other start-ups and third parties that complement them and supplement capability gaps.

French electrical engineering business Schneider Electric has, in its core business, always worked closely with technology partners on both hardware and software innovation. However, with the emergence of the digital era and more and more of their products becoming connected, the company realized it would need to open up to new partner businesses. These partners are often very different in culture, agility or maturity. Many are start-ups with only limited visibility on their long-term success or market uptake. The management at Schneider realized that it is for them to adapt rather than their partners. It therefore set up dedicated teams around the world in charge of developing relationships with these start-ups.[6]

Again, user-centricity is key. So-called 'use cases' are vital when working with young firms. You need to be very focused on and specific about defining the client issues you want to solve. Less is more when outlining your ideas.

2 Agile engineering in the New

Agile engineering goes mainstream

Agile engineering is another crucial capability going mainstream in the era of smart connected products. It is needed to continuously inject improvement into digitally and software-driven devices. This is a concept that was developed in the software sector but it is now being adopted more and more by hardware-producing industries. This should come as no surprise given the increasing software orientation of hardware we have already described. Product users increasingly demand permanent innovation and maximum adaptability from their devices for the sake of user-friendliness.

Against this backdrop, hardware manufacturers can no longer afford to bank on inflexible processes that bring, for example, a new car model to market over a timeframe of several years. Instead, constant iterative overhaul of ideas, products and attached services is needed.

Agile engineering means that teams swiftly gather feedback, test, and upgrade the architecture for products, software components and, crucially, the user experience. Connected to that is the idea of a minimum viable product (MVP), a development procedure in which a new product is developed with sufficient features, functionalities and services to get to market as quickly as possible and then iterate quickly based on real-world usage insights. Supported by digital manufacturing tools such as laser cutters, 3D printers or digital twins, hardware engineers can now develop ideas while concurrently testing them with users. 'From the first day you think about a smart connected product, you don't want to separate development and engineering from product usage and ultimately the end-to-end experience it provides,' stresses Olivier Ribet of Dassault Systèmes.[7]

Agile engineering also splits big projects into manageable units of work. It is essentially a form of micro trial and error or a chain of rapid prototyping cycles in order to get more guidance from real

Figure 9.2   Agile product development approach

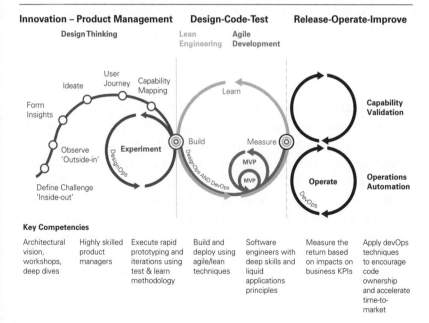

Innovation – Product Management		Design-Code-Test		Release-Operate-Improve
Design Thinking		Lean Engineering	Agile Development	

Key Competencies

Architectural vision, workshops, deep dives	Highly skilled product managers	Execute rapid prototyping and iterations using test & learn methodology	Build and deploy using agile/lean techniques	Software engineers with deep skills and liquid applications principles	Measure the return based on impacts on business KPIs	Apply devOps techniques to encourage code ownership and accelerate time-to-market

Figure 9.3 Engineering in the New – six key levers

6. Engineering Optimization
- Creating agility in R&D through operating model redesign, CoE structures, resource fungibility and optimizing core but non-differentiating processes
- Optimizing the product development process enterprise wide through value streaming and business intelligence

1. Modern Product Management
- Crafting go-to-market & product launch strategies, optimized product portfolios & product specifications and acting as product owners
- Designing transformational business model pivots (eg to aas) and data monetization strategies

5. Intelligent Product Operations
- Ongoing support and execution for NPI and PLM activities to give engineers more time to innovate
- Enabling technologies in predictive analytics, machine learning and robotic process automation to optimize engineering and product development

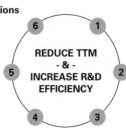

REDUCE TTM
- & -
INCREASE R&D
EFFICIENCY

2. Lean Agile Processes
- Improve efficiency and effectiveness of product development through lean agile engineering and 'shift left' processes
- Supporting these new processes with design thinking, continuous delivery and DevSecOps

4. Digital Thread & Twin
- Weaving the end-to-end product data continuity from design and collaborative development through manufacturing and support
- Leveraging a virtual representation to model, predict and provide insights on product behaviour from design through in-service

3. Connected Platform Engineering
- Modern SaaS and platform design, build and operate at scale, including end-to-end developer ecosystem operations
- Software modernization, refactoring and technical debt reduction to create product agility and market responsiveness

market feedback – and, through that, more accuracy in framing an intended user experience. Agile engineering teams are usually as self-governed as possible, executing two-week development sprints. Feedback loops guide the developer groups to create items that address client issues.

The developer groups, ideally a mix of designers, engineers, manufacturing experts, marketing, sales and service professionals, work as one collaborative team to develop rapid prototypes, built on cross-functional understanding. Learning from initial prototype testing and applying these lessons learned to define and develop the minimal viable product based on quicker iterations minimizes the conflict of functional objectives and results in improved speed of innovation while also significantly reducing the overall costs of development.

SMART VOICES   David Rush, Agile Engineering Lead, Accenture

Engineering in the New in the auto industry

The increased complexity of products, the increased need for innovation speed, and the increasing compliance requirements, mean many manufacturers are being forced to adopt 'Engineering in the New' approaches to unify hardware and software.

Car manufacturers in particular need an approach that balances agility and new engineering techniques with compliance. It is a process we call automotive lean-agile product development, with methods, tools and techniques including:

- scaling lean-agile methods to large, globally distributed product development teams;
- embedding systems and software compliance standards into product development process and methods;
- adopting modern engineering techniques to handle increased product complexity.

Accenture, working closely with automotive clients, has developed an industry-specific framework called 'AutoScrum' which combines an automotive-specific lean-agile engineering approach with a model based on lean-agile product development reference architecture. Its objective is to support and harmonize the software industry's agile working methods with the automotive industry's business model market cycle, and regulatory and compliance needs. Key attributes include:

Figure 9.4 AutoScrum in the New

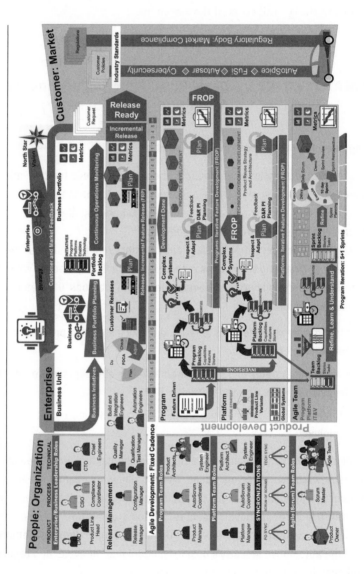

- Agile at scale
  - Fixed cadence.
  - Develop on cadence, release on demand.
  - System-feature-driven development.
  - Large-scale team collaboration.
  - Synchronized cross-disciplinary work.
- Modern engineering
  - Systems engineering discipline.
  - Model-based engineering.
  - Platform-based engineering.
  - Product-line engineering.

The AutoScrum framework is enabled by a prescriptive and unified ALM–PLM architecture that converges all product development work teams into a common language and collaborative ecosystem. The benefits of the AutoScrum approach are compelling, but significant organizational and cultural challenges need to be overcome:

- Companies push back on agile approaches as not suitable for hardware and complex systems and stay locked in traditional project paradigms.
- Companies resist using digital and model-based engineering techniques, remaining in a highly document- and text-centric environment.
- Companies struggle with reorganizing teams into core systems, platforms or domains.

Agile engineering requires new skills

On the one hand, the proactive character of smart connected products requires all engineering skills to work together on making the product collaborative and responsive. On the other, special expertise is needed to embed artificial intelligence.

These skill sets must be arranged into flexible and adaptable engineering team processes to adequately cover the complexity of integrated hardware and software development. This new engineering style needs a collaborative working mode between teams dealing simultaneously with hardware and software components – often spread across different regions and time zones.

Comprehensive software engineering skills are necessary to initiate a 'DevOps'-based process in which developers and operational staff, dealing with the product in action, collaborate. It means understanding cyber-security challenges and their implications for product and service engineering. Finally, it necessitates engineering the product as a digital twin, as this alone creates enough speed, flexibility and agility.

Smart connected products learn from their interactions with users and other products. This forms a basis for the delivery of customer-centric outcomes. For this purpose, product-related usage data needs to be generated and assessed. From an engineering point of view, this requires skills in digital and sensor technology, miniaturization, smart integration, electronics, wireless connectivity, cloud-based data storage solutions, and strong analytics and data science skills, all in order to transform data insights into meaningful action and product improvement. It furthermore requires skills such as experience designers, plus more traditional IT teams who can integrate smart products into enterprise systems. And in addition to the engineering skills around the product's collaborative attributes, there is a strong need for platform engineering. This includes among other things taking decisions on whether to create a proprietary platform or to leverage an existing one. All the skills and competences mentioned need to collaborate as one team.

Hardware is not dead

While we have argued extensively that value is moving from mechanical features to software and digital intelligence, this absolutely does not mean that physical hardware design becomes irrelevant. On the contrary, the best smart connected products still require truly great hardware engineering. Successful hardware engineers today combine classic skills like user ergonomics, usage of new and adaptive high-tech materials and fabrics, and miniaturization, with newer skills like agile iteration using 3D printing. These hardware engineers also need to have a much deeper understanding of and appreciation for both the data that will be generated by their hardware and the software that will enable the digital intelligence of the overall product. If we had to name a leading example we would go for Apple, where hardware engineers seem to have combined such skills in developing the groundbreaking, disruptive iPhone.

3 Data augmentation, leveraging AI

Smart connected products are data-driven and data-producing devices. Your organization must therefore be data-augmented in all its functions. Data should eventually become the main and most valuable currency within all your business processes.

Data is important for understanding the market, the usage of your product and ways to improve the user experience via new services. It will fuel the tools you'll use such as digital twins and will form the insights along your digital thread. It is – especially when AI-based analytics come into play – the new fuel in the New.

Data needs to be managed

Especially when AI comes into play, historic data can be used to train an algorithm. This algorithm will then be enabled to predict something – the computer model of a weather service can use yesterday's data to predict today's conditions. Once you have such algorithms trained, you need to feed them with operational usage data reported back from your products.

AI can also help to refine data and improve its quality. Structured data can be separated from unstructured data for further processing. In a wider operational sense this depends on creating a data model by which an organization defines what kind of data will be recorded and how it will be stored, processed and accessed. Often data of different unstructured formats and sources has to be harmonized into structured data sets. Most digital information needs such thorough rehashing before actionable insights on usage and user preferences emerge from product data. To choose the right analytics tools is therefore paramount. Data can also be commercialized in various ways. It can be sold directly to partners to work with it, be sold with matching data tools, be embedded into existing products and services, feed marketing blitzes, provide access to platforms for third parties, and help improve internal business processes.[8]

There is also a possibility that data products can hamper your existing core business because selling data often means selling a service. If you are a hardware manufacturer based on turnover from X number of sales per month and you move towards selling services, it is worth thinking about how to manage this shift in your business model without damaging your economic bottom line.

Rearchitect your data model

The data models used by the vast majority of product companies are decades old, and fundamentally ill-equipped for the smart connected product world. Historically, product companies defined a stock keeping unit (SKU) and each of these had a unique hierarchical bill of materials (BOM). This worked fine for decades when a product was simply sold to a distributor and fundamentally forgotten.

But this model completely breaks down with a product that is connected, upgradeable, customizable, personalized and perhaps sold as a service. This new world of living products requires a 'unified data model' that provides the flexibility and agility that is simply not possible with traditional data models. In many cases, the ideal data model moves away from physical components and becomes attribute-based. This so-called 'attribute-based data model' combines aspects of hardware, software and service-based models and provides increased flexibility to the product maker, the customer and the supply chain and ecosystem partners.

As shown in Figure 9.5, a unified data model expands vertically beyond the traditional hardware data to now also incorporate software and experience information based on customer-specific usage and personalization. The model also expands horizontally to ensure a consistent data model across engineering, manufacturing, sales and support.

Figure 9.5   Unified product model for capturing end-to-end business value

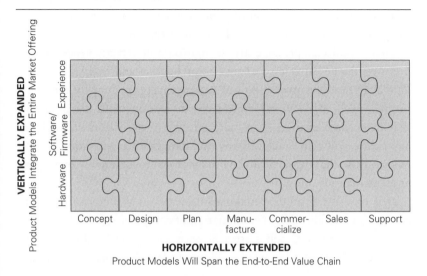

VERTICALLY EXPANDED
Product Models Integrate the Entire Market Offering

Software/Firmware Experience

Hardware

Concept   Design   Plan   Manu-   Commer-   Sales   Support
facture   cialize

HORIZONTALLY EXTENDED
Product Models Will Span the End-to-End Value Chain

Today most companies are far, far away from such a unified model. In a recent study, over 50 per cent of product companies reported having over 20 different product data systems. The reality is that current data models are very siloed and need to be completely reinvented.

Choose the right data tools

The analytics tools dealing with your data have to match your ambitions. Always think about involving AI as this pushes your data insights in most cases into a higher quality league. For instance, a leading semiconductor company is using advanced machine learning and AI tools to schedule engineering simulations and accurately predict the engineering compute and storage requirements. Other product companies are applying machine learning techniques to the product introduction stage to conduct manufacturability assessments.

Once again, the digital twin and digital thread concepts are very handy in this context in that they incorporate all of this, acting as a complete data representation of your product in action, giving you the full picture on usage, insights and proposed actions.

Smart connected products are strongly data-driven and hence companies need to be capable to apply comprehensive and multidisciplinary data analytics. The involvement of AI is not just necessary for the smart product itself, it is also needed to generate learning software programs that make data analysis even more effective. Such AI-driven analytics algorithms will help to identify relevant data findings more quickly and efficiently, so that big data can be broken down to actionable smart data at greater speed.

It is crucial for product-making businesses in this context to build data-augmentation skills, a combination of computer science and IT expertise, mathematical and statistical know-how, and product-relevant domain and business knowledge. However, data augmentation should always target intelligent customer-centric outcomes and not be done purely for its own sake. To ensure this, a company needs to define principles for data augmentation that keep it open-minded about, but not overrun by new sources of customer value across the product lifecycle. This requires the right talent – data scientists – currently one of the most in-demand human resources globally. Companies need to decide their 'make vs buy' strategy in data and AI services; specifically, whether to aggressively build this talent internally or purchase these skills from ecosystems partners.

4 'As-a-service' competencies

As highlighted in Chapter 5, the shift from selling hardware to a recurring as-a-service model is indeed a very challenging one. Most product companies struggle greatly with it as it requires a fundamental change to their very DNA. Moreover, this requires building a wide array of capabilities that do not exist today. Figure 9.6 shows a product-as-a-service capability framework that presents at a glance the manifold points to keep in mind when leveraging smart connected products to transition to an as-a-service business model.

Figure 9.6   Product as-a-service capability framework

Define	Acquire	Build	Manage	Support
Strategy	Sell	Development & Testing	Finance & Legal	Customer Sales & Service
Business Model Strategy	Marketing Strategy	Security & Regulatory Compliance	Revenue Recognition	Customer Success
Experience Design	Customer Targeting	IT Operations	Reporting & Compliance	Adoption & Consumption
Competitive Strategy	Configure, Price & Quote	Device Development & Testing	Accounting	Entitlement Management
Go-to-Market	Asset Deployment	Research & Development	Contracts	Service & Support
Organization & Culture	Kitting & Packaging	Methodology & Tools	Invoicing Process/ Management	Device Management & Maintenance
Governance	Install & Test		Device Management & Maintenance	Repairs & Warranty
	Repair & Reverse Logistics		Security Policies	End-of-Life Services
	Product Operations		Device Configuration	Service Levels
	Order & Supplier Management			
	Global Sourcing & Partnership Models			

Within this overall framework, many of the capabilities have been discussed elsewhere in this book, but here are some additional specific skills that most product companies struggle with today as they attempt the shift.

Solution configuration, pricing and quoting (CPQ). An 'as-a-service' model offers customers a higher degree of choices and customization options than a typical hardware product sale. Most product companies do not have processes or IT systems to enable these more complex configuration, pricing and quoting (CPQ) capabilities.

Sales. Traditional product sales forces are trained and incented to sell a product and then move on to the next sale, which fosters a very transactional mindset. Service-based models do not work this way. Instead, sales personnel must think more about customer usage and adoption of the solution over the lifecycle. Indeed, many service-based companies actually defer sales commissions to be paid out over the lifecycle of the service and create bonuses based on customer usage. Figure 9.7 illustrates the sales transformation capabilities needed and the current maturity status in several industry segments.

Service assurance. An as-a-service model typically comes with service level agreements (SLAs), which are commitments over the entire lifecycle. Most product companies have no processes or systems for even measuring SLAs, much less taking corrective actions when SLAs are not meeting targets.

Customer success. In most as-a-service models, the revenue received is linked to customer adoption and usage of capabilities. The software industry, which we pointed out earlier is leading in the shift to as a service, has pioneered the creation of an entirely new function called 'Customer Success', which is staffed by experts who deeply understand the customer's organization and processes and are measured on customer adoption metrics. This is a completely foreign concept to most hardware companies.

Entitlements management. In an as-a-service world it is essential to understand the rights or 'entitlements' at the individual device and user level. For example, which devices are entitled to receive an upgrade and which are not, or which users are allowed to access premium features versus only the basic features. This requires sophisticated entitlements management systems that virtually no product company has in place today.

Figure 9.7   **Sales transformation of industrial equipment companies**

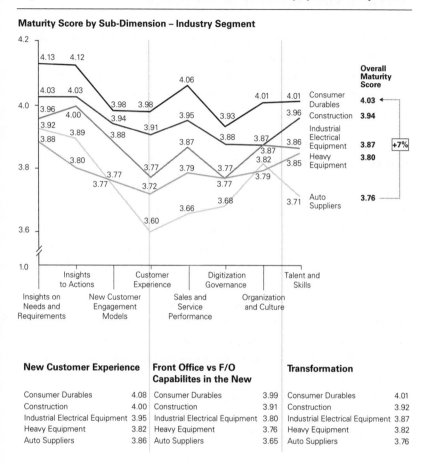

Maturity Score by Sub-Dimension – Industry Segment

			Overall Maturity Score
Consumer Durables			4.03
Construction			3.94
Industrial Electrical Equipment			3.87
Heavy Equipment			3.80
Auto Suppliers			3.76

+7%

Data points by segment:

Insights to Actions (Insights on Needs and Requirements): 4.13, 4.12, 4.03, 4.03, 3.96, 3.92, 3.88, 3.80, 3.77

Customer Experience (New Customer Engagement Models): 3.98, 3.98, 3.94, 4.00, 3.89, 3.88, 3.77, 3.77, 3.72, 3.60

Digitization Governance (Sales and Service Performance): 4.06, 3.95, 3.93, 3.91, 3.87, 3.88, 3.79, 3.77, 3.77, 3.66, 3.68

Talent and Skills (Organization and Culture): 4.01, 4.01, 3.96, 3.87, 3.87, 3.82, 3.86, 3.85, 3.79, 3.71

New Customer Experience		**Front Office vs F/O Capabilites in the New**		**Transformation**	
Consumer Durables	4.08	Consumer Durables	3.99	Consumer Durables	4.01
Construction	4.00	Construction	3.91	Construction	3.92
Industrial Electrical Equipment	3.95	Industrial Electrical Equipment	3.80	Industrial Electrical Equipment	3.87
Heavy Equipment	3.82	Heavy Equipment	3.76	Heavy Equipment	3.82
Auto Suppliers	3.86	Auto Suppliers	3.65	Auto Suppliers	3.76

Infrastructure support. Smart connected products need to be monitored, and they generate a lot of data that needs to be securely stored. Most product companies lack the IT infrastructure required to manage as-a-service installed-base analytics. In addition to securely storing the data generated by the fleet of smart connected devices, the much harder skill is deriving insights from it and making recommendations for what a customer should do. This requires not only data scientists but also personnel who understand the customer's industry and business processes.

5 The experiential workforce

A workforce dealing with the creation of smart connected products needs special skills, profiles, mindsets and behaviour. An outcome-oriented talent pool is a prerequisite, because the New is all about the development of devices meant to provide intelligent customer-centric outcomes along the entire product or service lifecycle.

Build a customer-centric organization

Enable your workforce to think and act along end-to-end experiences and outcomes instead of mere product features and outputs. Executive teams need to be guided by new incentive metrics, shifting from units sold to customer success achieved. An ecosystem and new talent from outside companies, which already sell outcomes, will support this workforce transformation.

The shift could be complemented by hired talent, which already has the outcome-oriented mindset. The necessary inspiration for that shift should be driven by a CEO-defined vision of outcome-driven smart connected products, an experience-driven ecosystem, and the respective ecosystem partners.

To infuse the inspiration into the workforce, top management must lead, convincing personnel of the success of a smart connected product strategy, transforming them from output-oriented to outcome-oriented.

Build a fluid, agile, innovation-led organization

Leaders at every level have to change their mindset to thrive in the digital age, in exponential organizations and agile working environments. The traditional business leader was in the past rewarded, with promotions or improved financial compensation, when he or she could maintain control, stability, structure and consistency in a business, when individual performance and expertise lived up to expectations, and when market power could be built for the organization.

Figure 9.8 New leadership is all over

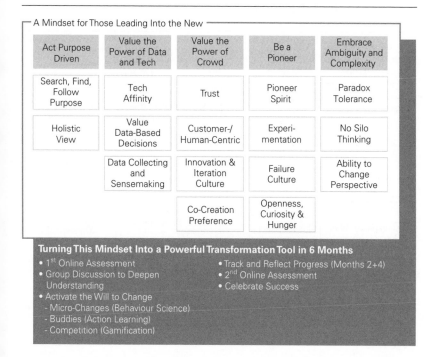

A Mindset for Those Leading Into the New

Act Purpose Driven	Value the Power of Data and Tech	Value the Power of Crowd	Be a Pioneer	Embrace Ambiguity and Complexity
Search, Find, Follow Purpose	Tech Affinity	Trust	Pioneer Spirit	Paradox Tolerance
Holistic View	Value Data-Based Decisions	Customer-/Human-Centric	Experimentation	No Silo Thinking
	Data Collecting and Sensemaking	Innovation & Iteration Culture	Failure Culture	Ability to Change Perspective
		Co-Creation Preference	Openness, Curiosity & Hunger	

Turning This Mindset Into a Powerful Transformation Tool in 6 Months
- 1st Online Assessment
- Group Discussion to Deepen Understanding
- Activate the Will to Change
  - Micro-Changes (Behaviour Science)
  - Buddies (Action Learning)
  - Competition (Gamification)
- Track and Reflect Progress (Months 2+4)
- 2nd Online Assessment
- Celebrate Success

In the new product world things are rather different. Leaders will be judged against their capacity to be customer- or employee-centric, whether they show open-mindedness towards new ideas, manage to push agile work styles and have more technological understanding compared to former executive generations. This needs a new kind of management talent whose components are broken down in Figure 9.8.

This is where the exponential organization comes into play. To bring exponential organizations to life requires diverse, complementary teams and smart systemic steering. But once this has been achieved they can be 10 times better, faster and cost-effective compared to their rivals. They have then also found a way to leverage organizational, structural or technological innovations that were inconceivable before the digital age.

Figure 9.9 on page 200 shows a typical set of arguments heard from business leaders that hold them back from becoming Exponential Organizations (ExOs).

Figure 9.9   Most common barriers hindering a company's exponential growth journey

We are not a start-up. These new ways are not suitable for our tradition.	Why should we change? We are (still) successful.	Management does not really want 'the New'. They keep rewarding traditional behaviour.

Lack of orientation: What is in it for me?	We don't have the right people.	Experimentation is a career stopper.	We are afraid of the unknown.

We are lacking the relevant skills.	Old control mechanism	No employee empowerment

Existing decision-making processes hinder innovation.	Agile will not work here. We are different.	Failure is not an option. Experiments and MVPs are not acceptable.	Shareholder Alignment

Rigid IT Infrastructure	Employees and managers believe the owners/stake-holders don't want the New.	Lack of available data and systems in the first place.

Human + machine: how to best collaborate with your AI colleague

Today's workforces are becoming more and more mixed. Robots and cobots have become sophisticated tools that can help humans accomplish a variety of high-value tasks, and their capabilities are progressing. We foresee a future where humans and machines will increasingly collaborate seamlessly.

The latest generation of robots includes machines that are more mobile, flexible, autonomous and affordable. As these robots' intelligence and autonomy increases, worker perceptions of them will shift from them being tools to being teammates. Industrial tasks will increasingly be accomplished by coordinated human-robot teams. Each of the teammates will have their own tasks to accomplish and, increasingly, those robotic systems and their human co-workers will need to negotiate and cooperate on their own distinct yet inter-independent tasks in order to accomplish sophisticated activities more effectively and efficiently than either humans or robots could do alone. Robots will be controlled via a combination of pre-defined automated plans and processes, but ad-hoc improvisation will be required when a human teammate sees that things are not going as planned.

Imagine a robot that can navigate in factory- or warehouse-type environments, and either scout for problems or transport objects. A shipment of goods and material has arrived, and both connected human workers and a set of transport robots are divvying up and distributing the items to appropriate places.

Both individuals and teams – human teams, human-robot teams, and robot-robot teams – will be executing these tasks. As they work, additional interactions will arise between those humans and robots – and between the other people in the building. For instance, perhaps a person who has just received some material will want to co-opt the robot deliverer's help for half an hour. Perhaps there's an urgent need to grab an additional, unanticipated item. Or perhaps another agent, a robot or connected human worker, will have been called away, necessitating a replacement. Smooth and effective execution of complex processes in situations such as this, by human-robot teams, will soon be mainstream.

6 Ecosystem orchestration[9]

Not all, but some smart connected products will work as platforms, as we have shown in Chapter 6. In order to create and run a successful platform it is essential to have the capabilities needed to build and orchestrate ecosystems that support such product-based platforms. These ecosystems must then be capable of identifying potential partners who can contribute crucial technologies, data or service elements to boost the value of the product.

Only such partner ecosystems will fuel breakthrough innovation and disruptive growth opportunities. Operating models built around such multilateral and multidimensional partnerships and open collaborative networks will also create sufficient value to secure survival in the new product world. Michael G Jacobides, Sir Donald Gordon Chair of Entrepreneurship and Innovation at the London Business School, states:

> 'I think that the rise of the smart connected product is part and parcel with the development of seamless ecosystems that make life less stressful for users and more customized to the needs of the individual. Most smart products would probably not yield financial successes without being woven into an ecosystem.'[10]

Ecosystems are often open networks of strategic business partners with the common aim of driving growth and fostering innovation. It can be seen as a company's competitiveness network forming as an increasingly global, data-driven, multi-industry cocoon around its participating partners – consisting of cooperators, suppliers, institutions, customers and other stakeholders.

All these players connect and collaborate in order to innovate more quickly, to amalgamate complementary capabilities for better results and to react with greater agility to fast-moving consumer and business markets that increasingly demand complex outcomes rather than monodimensional products.

The ecosystem is going to be a form of life 'tissue' around you, without which you will not have the muscle to stay in business. Typically, they are built around intellectual property stemming from one core partner, which is combined with service hardware and various APIs and offered on an as-a-service basis to a horizontal or vertical market.

Industrial businesses should first and foremost aim to create these with viable networks of partners with the shared aim of developing new products and services targeted to outcome-focused customers. In a second step they can then – but need not necessarily – adopt a strategy where their products are more and more shaped as platform items so that they can serve as the nucleus further down the line for a highly value-creating product-centric ecosystem.

From a managerial standpoint, this will be complex, as industrial businesses will have to create agile global organizations with a pragmatic sense for bonding, which are much more permeable, flexible and inherently collaborative. That will require changes throughout the organization – in people, technology and strategy.

Business leaders must therefore set the tone and stage by declaring a shift away from products and services alone towards how the company can help customers achieve better outcomes. To lay the foundations, they must recognize that employees at all levels have connections that could serve the company's ecosystem strategy.

Figure 9.10 shows a list of steps to make when it comes to creating highly effective ecosystems and using them as an innovation hub.

Figure 9.10 Creating highly effective ecosystems

Set the Direction	• Clear definition of a 4–5-year future state is critical to the success of the overall ecosystem
Executive Sponsorship	• Strong, committed executive sponsorship of the governance structure • Process that allows for quick decision making and issue resolution
A Hands-on Steering Committee	• The governance structure should help with the ecosystem strategic partnership strategy, eg identifying and facilitating other ecosystem participants
Diversify KPIs	• Define metrics and KPIs that measure the success of the ecosystem strategic partnership over time; do not rely on sales numbers only
Channels, Channels, Channels	• Activate the channels from the very top; have a holistic approach to a client problem, industry channels, and operational group
Solution Roadmap	• Do not build a large portfolio of solutions; validate hypotheses with clients, industry Subject Matter Experts (SMEs), etc before development; refine before investing in full on development to increase solution ROI • Consider customer proof of concepts a solution as part of the contract
Manage Global Scope	• Don't spread resources too thin too early • Prioritize and manage geographic scope with global support teams
Early Wins/Intercept Inflight Opportunities	• Secure early wins to build and hold momentum and 'mindshare' within each organization
Flexible Operating Model	• Define a flexible operating model to remain responsive to market • Ongoing joint innovation performance tracking and program management process
Change Management/ Cultural Alignment	• 'One team' philosophy (process, metrics, attitude) • Treat the ecosystem strategic partnership like a newly created business unit • A vision that allows the ecosystem strategic partnership to focus on high-value activities that will drive results

To be able to assess the right agreement with partners is important – not least to arrive at the right formula of value sharing within an ecosystem. The platform concept means profit sharing of some kind. 'All of a sudden you have that reallocation of value amongst the different players in an ecosystem around a smart connected product. With that reallocation of value the core functions of a business can shift', says a former Tesla manager we talked to.

An understanding must be developed of which platform partner deserves which share of revenue to create a margin from its involvement. In many cases, partners will also be competitors in other areas, so the right system of Chinese walls will need to be put in place. For the management of an ecosystem, a new company function should be created. This requires a high level of expertise in managing the details as they stand and understanding how such business models could evolve over time.

7 Pervasive security

Smart products are programmed to perform certain actions that often transcend cyberspace and impact the physical world, meaning safety becomes a major concern. Managing these devices demands the capability to define and manage a broad range of security aspects. Within the ecosystem, cyber security risks must be anticipated and proactively managed, and intellectual property (IP) must be protected.

A central prerequisite for this ecosystem is to partner with the necessary expertise to provide relevant components to the value creation mechanism of smart connected products. Security and IP protection must be thought through in detail at component levels. 'You should put data security above everything to avoid damage done through attacks to the safety and privacy of customers. Malevolent hacks have become a not-too-uncommon occurrence in the world of connected products', says the former Tesla manager.[11]

Businesses should aim to define security standards that are mandatory for all involved ecosystem partners. But they should also leverage the power of the ecosystem in order to cross-pollinate on security-relevant solutions for the product. For this purpose, data sharing principles for the ecosystem must be defined and security standards made visible and transparent to all partners.

The ability to maintain and protect sensitive data and ensure that only authorized entities have access to it is vital. This applies both to data at rest and data in transit during communications. But similarly vital is the ability to sustain operational continuity. Data, services, networks and applications should be accessible in a timely manner when needed.

Still, there remains a string of security risks lurking when smart connected products start to play a role in business models. Here are the most important to keep in mind.

Unlike many traditional devices, smart connected Internet of Things (IoT) devices, being software-updated, are expected to have long lives. Keeping smart connected devices up to date with the latest security upgrades is essential. But applying updates is not always possible, for example because the high availability requirements prohibit or the outdated hardware is not capable of receiving the requisite patches. This leaves devices vulnerable to exploitation in the later stages of their life.

The IoT has also introduced many new communication protocols and connectivity options. But a lack of standardization coupled with limited security awareness by device manufacturers has led to inconsistencies and grey areas in their implementation and usage that can be exploited by cyber attackers. Furthermore, IoT markets are extremely price competitive. That can mean low-cost components are fitted with lower security standards. Devices could therefore have insufficient hardware resources available for basic security functionality such as encryption, password management, or secure storage.

The autonomous operation of machines means that often there is no human present to authenticate a device's identity. But establishing automated relationships between machines based on conventional identity mechanisms remains, from a technological standpoint, incredibly difficult. Finally, the IoT by nature bridges previously disconnected domains, significantly blurring the borders of what traditional regulations cover. In this evolving landscape, even staying compliant with changing regulations becomes challenging.

Takeaways

1 To successfully reinvent your product, seven pivotal capabilities must be built.

2 While many of these new capabilities focus on the product development function, the shift to as-a-service business models affects virtually all processes and organizations.

3 In addition to building the new capabilities, a fundamental cultural and mindset change is required.

10

The roadmap to success with living products and services

CHAPTER SUMMARY

....

Successfully repositioning a business towards smart connected products might seem a protracted maze to many leaders of today's product makers. However, it is important to quickly embark on this journey and we have defined a structured list of action points with clear markers to guide you. It requires an open entrepreneurial spirit and a shift in mindset, from envisioning features to envisioning experience. This in turn implies a new product execution roadmap and new business set-up processes, with siloes broken down, new roles and responsibilities added and the IT landscape re-architected. Once in operation, the new structures and processes around developing, making, shipping and field-managing smart connected products need constant monitoring and tactical adaptation via the data the products feed back.

....

10

The roadmap to success with living products and services

CHAPTER SUMMARY

Digital is now everywhere – almost. Actually, far too often, it is still a remote and abstract idea in the boardrooms of many product-making companies, in both business-to-business and business-to-consumer markets.

In our day-to-day client advisory work, we still meet numerous executives and line managers who envision value-driving products predominantly in conventional categories. There is only limited scope in their plans for things like software-driven smart capabilities, or service-oriented user experience, let alone platform aspirations. There is, generally, a strong failure to perceive both the enormous value potential of smart connected products, and the massive downside risk if a competitor or new entrant successfully disrupts your industry by making their product smart and connected first.

In 2018 we ran a global survey of almost 1,000 senior business executives. While 68 per cent said that they believed in the necessity of pivoting to a future of smart connected products, only 16 per cent confirmed that they were really envisioning such a step and even fewer said they had a clear roadmap.[1] So, the rift between theory and practice on this is huge among business leaders, presumably because these new sources of income are not easy to spot and the actions required are unclear. To render products smart and connected, it appears, is still all too often seen as a nice-to-have extravagance rather than an economically sound expansion move into emerging markets just waiting to be tapped.

We asked James E Heppelmann, founder and CEO of PTC, a software company catering to various industrial sectors, for his take on the findings. To this widely respected commentator on the industrial sector's digital transformation the explanation was simple: 'For someone who grew up surrounded by physical products all these new software implications can be a scary place'.[2]

So how can top executives overcome their fear of reinventing product lines? What will dispel the anxiety and scepticism about adding smart characteristics to a product portfolio? What is needed is a map of all this so-far uncharted territory.

To that end, we present seven key action points for pivoting to smart connected products. These should give you a clear roadmap and also help inspire your organization with the necessary belief, vision and determination.

Marker 1: Definition of vision and value spaces

The journey starts with believing in the economic superiority of smart connected products, understanding that they push open doors to entirely new value perspectives – self-feeding platforms, agile, lucrative ecosystem alliances, and numerous profitable services previous generations of product makers could never have dreamed of delivering. From there, you must rapidly envision these possibilities in ever more concrete designs and blueprints so as to be able to execute swiftly and successfully.

An entire organization will not be truly convinced by this new world overnight. But a good first step is the CEO level consciously focusing on a future in which the key to a product's success is not being feature-driven or even just connected but driving hyper-personalized experiences through intelligent adaption. This means business leaders imagining their pure hardware products becoming increasingly intelligent over time. Products become containers for software and digital intelligence updated and controlled remotely, allowing customers to customize and personalize experiences as they wish.

To make its vision compelling for employees, customers, suppliers and ecosystem partners, the company must define the type of digital product it wants to build, either from scratch or by reinventing items from its existing line. To help guide this decision, let's revisit the Product Reinvention Grid discussed in Chapter 3, which offsets a smart product's level of technological advancement – its Intelligence Quotient (IQ) – with the level of user experience quality it delivers – its Experience Quotient (EQ). Building on this framework, we have identified five specific 'value spaces' where we believe profits are most likely to be found, as shown in Figure 10.1.

Figure 10.1   Five value spaces for product companies

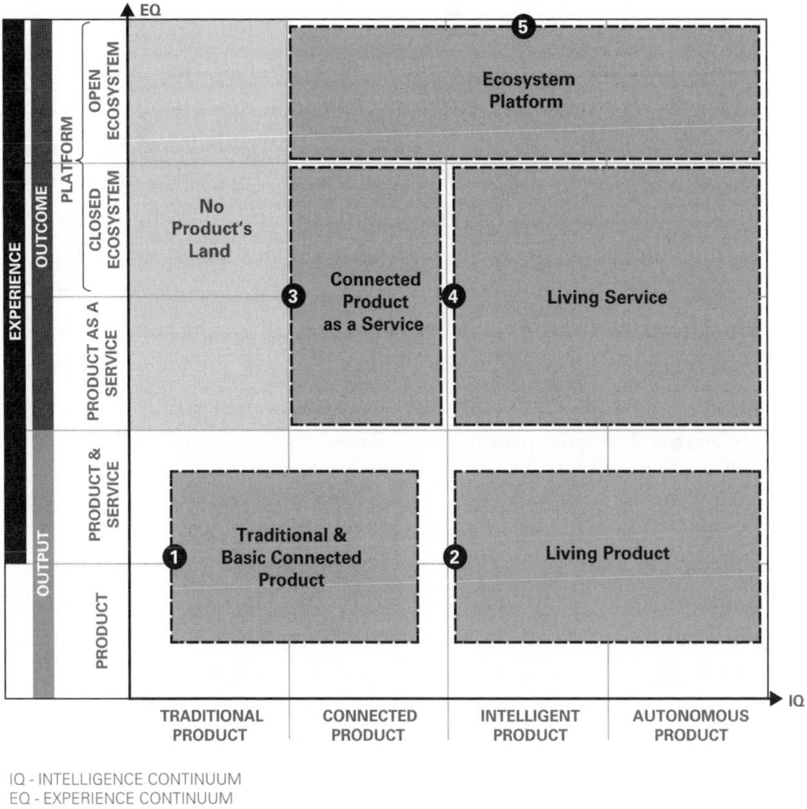

The ideal value spaces for your company or your specific product will vary. There is no one-size-fits-all solution, but we believe there are some standard guidelines to follow.

The value spaces are as follows:

1 **Traditional product and basic connected product.** This is the value space that most products occupy today. Our belief is that very few products will remain profitable in this space over time. The vast majority will need to be reinvented to move up the IQ and/or the EQ axis to a new value space. Steve Myers, the CEO of Mindtribe,

recently hosted a meeting with several product innovation luminaries in Silicon Valley who agreed with our conclusion that 'the traditional product business model is dying'.[3]

2 **Living product.** A living product is a product that leverages AI technologies extensively, either as embedded functionality or via an edge or cloud network. As we have previously described, the rapidly falling costs of sensors and connectivity coupled with growing accessibility and acceptability of big data and AI technologies will fuel a tremendous growth in offerings in this value space. Indeed, Rajen Sheth, Senior Director of Product Management for Artificial Intelligence at Google Cloud, believes that 'all products will utilize AI within the next decade'.[4] We believe this will be a very profitable value space for many product companies. Moreover, we see no alternative for most product companies than to increasingly move to the right on our IQ axis.

3 **Connected product as a service.** As we said in Chapter 5, we strongly believe that many product companies will migrate from transactional hardware to a recurring as-a-service business model. As evidenced by the Adobe story from the software industry, there is unquestionably significant value to be unlocked by moving to this model.[5] Our view is that most companies making connected products should immediately explore a movement up our EQ axis to develop new as-a-service business models as a first step.

4 **Living service.** This value space combines the previous two: a living product with an as-a-service business model. A living service, as the name implies, is a more dynamic, adaptive service that evolves over time. We anticipate a tremendous amount of dynamism and innovation in this space, as AI will unleash a set of customizable and personalizable services that would have been simply impossible only a few years ago. As living services deliver enhanced value for the individual users/customers, our belief is they will also prove a very attractive value space for makers.

5 **Ecosystem platform.** As described in Chapter 6, platform business models are creating tremendous market value, so this is clearly a profitable value space. The core question for a product company is: will you fundamentally change your business model to become a platform as the main basis for revenue generation or will you simply leverage a third-party platform to deliver your living product or living service capabilities? This is not a nuanced question. We believe very few product companies can or should aspire to this value space, despite the potential rewards.

The key strategic choice points for any product company are first, how fast and how far to move on the IQ axis towards a living product and second, whether to shift your business model up the EQ axis from transactional product sale to as-a-service or ecosystem platform.

In our experience, investors and employees are demanding that companies develop compelling visions of how not just their products but their internal business structure and processes will digitally evolve. To decide to invest in all this obviously requires entrepreneurial guts on the part of executives, but this courage must be a function of conviction. The vision must be backed by a strong rationale if the changes are to benefit the business.

Marker 2: Digitization of the core business to fund expansion

The shift from the current product business to a new value space such as living products or living services, will require substantial investments. The transformational path from old to new therefore needs to be charted and a business case spelled out in detail, with appropriate funding dimensions in mind. An economic master plan must capture how pivoting into the new value space will impact the company's finances over time. This is especially true when shifting up the EQ axis to as-a-service business models, as this transition will have a major impact on both the business balance sheet and the income statement.

Figure 10.2 The 'Rotation to the New' framework

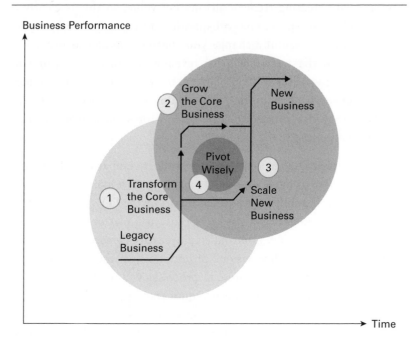

To fund the requisite investments and mitigate the risks, a holistic approach to managing the transformation is required. The 'Rotation to the New' framework shown in Figure 10.2 has proven a very effective model for structuring and guiding the transformation. The concept is that most companies have a very well-established 'core business', which for product companies is a decades-old model of selling traditional products as a transactional hardware sale. We strongly advocate that this core business must be grown but also complemented by the New, which for most means a move to living products that could also include living services. This difficult transition must be managed via a wise pivot that protects the core business while scaling the New.

Finding the monetary means for a change of such proportions is one of the tougher challenges because the transformation is not only of product lines but also, crucially, the whole business organization. Even if the executive board are convinced of the need to reinvent their products, they tend to remain tight-fisted when it comes to putting serious money behind the process because smart products can only function and unleash their value creation when they are made and managed under the auspices of new organizational principles. And this change seems to carry considerable risk.

The good news, our experience shows, is that the digital transformation of today's core business can drive massive cost savings while simultaneously building the necessary foundation of digital capabilities. Moreover, the efficiency gains transforming the core are quite immediate and can be measured using established, market-accepted metrics, so their advantages can be clearly perceived by top management and boards. Furthermore, the moves are sufficiently value-creating to transfer a surplus into the push for the new world of living products or Services. Indeed, our research shows that digital transformation programmes can save 300 to 700 operational cost basis points, which typically is more than enough to fund the requisite investments in the New.[6]

For an example of a traditional product company successfully making such a dramatic 'Rotation to the New', we can look again at the Haier Group, the Chinese company that is the world's largest domestic appliances maker. Chairman and CEO Zhang Ruimin outlines a new approach to business at Haier marked by three main characteristics, representing fundamental changes to traditional product makers. In the first move the enterprise was transformed from a closed to an open system. This shift happened via the introduction of a network of self-governing micro-enterprises with free-flowing lateral communication between them and mutually creative connections with outside contributors. In the second move, employees were transformed from being executors of hierarchical top-down directions to being self-motivated contributors, in many cases choosing or electing the leaders and members of their teams. And in the crucial third move, the purchasers of Haier's appliances were transformed, from the perspective of development and management teams, from traditional customers to lifetime users of products and services designed to solve their problems and improve their user experiences.[7]

Marker 3: Sketching the smart connected product roadmap

Once you plot your aspirational vision, a detailed roadmap is required to chart both your product intelligence evolution on the IQ axis and your experience evolution on the EQ axis. Evolution on both axes presents major challenges for traditional product companies, and a thorough understanding of the combinatorial effort required based on the Product Reinvention Quotient (PRQ) is critical.

Increasingly, product intelligence on the IQ axis requires a fundamental rethinking of the overall product architecture. Examples of capabilities that need to be incorporated include:

- Sensors: What additional sensors should be included to deliver on future desired experience or intelligence?
- Voice user interface (UI): Do you want a voice UI? If so, which voice platforms(s) will you support?
- Additional digital UI technologies: Should you add gesture-based controls? New generations of flexible touchscreen technologies?
- Security: What new security protections are embedded in your product?
- Communication: Which protocols will your product support (and not)?
- Open vs closed: How open or closed to third parties will your product be?
- Upgrade-ability: What components of the product will and will not be upgradeable?

Ideally, your product architecture should also be as future-proof as possible, given the rapid pace of technological advances, but this is no easy task. Future proofing comes at a cost. Tesla is a good example of a company that thinks ahead strategically about future desired capabilities and includes additional sensors and upgradeability to enable future improvements. For example, acceleration mode in Tesla's model S initially brought the car from 0 to 60 mph in 3.2 seconds. A simple, remote software update lowered that to 3.1 seconds.[8] It was a significant improvement of the car user experience and only doable with a connected and software-rich device. Another example was the upgrade to autonomous driving capabilities, which was only possible because Tesla installed additional sensors in their cars well in advance of their actual usage.[9] In traditional cars, even with as many computers as they now contain, the software would not be deeply embedded enough to influence such physical performance features.

Still, there is a point in keeping hardware up to date. In consumer products in particular, hardware is and will remain the first point of connection for users – despite ever-increasing amounts of software content inside. To upgrade hardware quickly and cost-efficiently, modular manufacturing is the way forward; parts and components are predefined and their future evolution path already charted as far as this is possible today.

The experience quotient (EQ) upgrades require consideration of when and how to upgrade to an as-a-service or ecosystem platform model. The ecosystem model requires a wide variety of collaborators – the vendors, service providers and other parties in the ecosystem. Think clearly about how interfaces can be adjusted from a technological standpoint along the usage patterns of the experiences delivered by the product. The final shape of your product should have a predefined Product Reinvention Quotient, so it is worth fixing the IQ and EQ levels in advance. Given that an intelligent product can communicate or connect with others, build in the possibility for users to also collaborate and help improve each other's experiences without bringing the manufacturer or vendor into the picture.

As the pace of both IQ and EQ innovation is so fast, your roadmap must be built on a foundation of agile methodology and follow the minimum viable product (MVP) principle in product development cycles.

Finally, irrespective of the goal the product is intended to deliver, it must act responsibly in the way it treats customers and their data. It must be designed, built and serviced to engage and not eavesdrop. The data it captures and that runs it must be secure and sharable only with consent. The product must remain resilient against malevolent hacks and be autonomously or manually updated with security features and firmware updates at regular intervals. Various ecosystem partners engaging with the product must be equally bound to protect customer privacy.

Marker 4: Creation of a digital innovation factory to accelerate the change

We cannot emphasize it strongly enough: building the skills and new organizational culture needed to succeed in this new digital product world is a simply enormous challenge. As discussed in prior chapters, each of the five big shifts requires capabilities not typically present in traditional product companies. For example, taking traditional unconnected hardware products to higher advancement levels – smart, intelligent or autonomous – requires a fundamental rethink of design and product development. Shifting to a platform-based product requires the building of new software platform engineering and ecosystem management skills.

Furthermore, a whole spectrum of new technology skills needs to be built, ranging from AI, edge intelligence and extended reality to 3D printing and advanced analytics. The new world compels engineers to ask hitherto unknown questions, such as: What level of product functionality should be embedded in the cloud versus in the product itself? How should network infrastructure be designed to drive quick and easy upgrades? How deeply should software be embedded into a hardware product?

Figure 10.3 Digital factory objectives

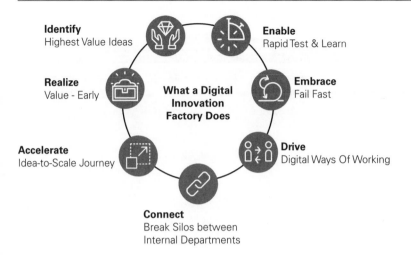

Identify
Highest Value Ideas

Enable
Rapid Test & Learn

Realize
Value - Early

What a Digital
Innovation
Factory Does

Embrace
Fail Fast

Accelerate
Idea-to-Scale Journey

Drive
Digital Ways Of Working

Connect
Break Silos between
Internal Departments

The more a product climbs up the ladder on the IQ or EQ axis in our new value chart, the more specialist knowledge is needed to help projects move forward. Data scientists, visualization experts, automation specialists, software platform engineers, and system architects are just a handful of the new roles. The downside is that these roles are often so specialized and held by such a tiny number of people that the salaries will be high. So a core question is, how can a traditional product company attract and retain the talent required to succeed?

One proven approach to succeeding in this new world is to create a digital innovation 'factory'. This is not a manufacturing factory, but rather a product and experience innovation centre housing all the interdisciplinary skills required. The focus is on designing and producing new digital products and services with dramatically faster innovation cycles using new design thinking and agile techniques. The innovation factory can also serve as a change agent vehicle for the entire company to embrace key concepts such as rapid test and learn and acceptance of fail fast. These concepts are laid out in Figure 10.4.

Figure 10.4   House of capabilities to support and accelerate digital transformation

HOUSE OF CAPABILITIES			
VISION, STRATEGY AND GOVERNANCE	START-UP/ PARTNERS SCREENING & EVALUATION	INTEGRATED SUITE OF PRODUCTS+/ SERVICES	ECOSYSTEM OF PARTNERS

IDEATION	INCUBATION PROTOTYPING	INDUSTRIALIZATION (BUSINESS & TECHNOLOGY)
Design Thinking Value Chain Analysis, Voice of Customer, Customer Journey, Ideation Workshop, Idea Qualification, Assessment & Prioritization	Design Value Proposition (CVP), Business Design & Operating Model, Rapid Prototyping: Design & Test of Mock-up/Proof of Concept	Eg Packaging, Go-to-Market, IS/IT, Billing Services, Installation, Offer Subscription

SENSORS, CAMERAS, ACTUATORS	APPLIED INTELLIGENCE (Data Acquisition, Algorithms...)	ARTIFICIAL INTELLIGENCE	REMOTE DATA LAKE (Private/Semi-Open)	SERVICES MONETIZATION PLATFORM
	Edge Applied Intelligence Computing ECU			
	On-Board Applied Intelligence	CYBER-SECURITY	API GATEWAYS TO 3RD PARTIES	BACK OFFICE INTEGRATION
	Off-Board Applied Intelligence			

Typical Staffing: 150–300 people for a US$20–30 bn company

While there are standard components and skills required for a digital innovation factory, there is no standard organizational model. Every company must create its own, based on its existing position within the product evolution space and its ambitions for the future. Initially, this factory could easily be an external entity created with partners to be insourced to the company at a later stage. For many product companies, the factory should also include a next-generation production shop bringing together a wide variety of interdisciplinary skills under one roof.

In our interview with him, Rich Lerz, the CEO of Nytec, shared his experience of developing a digital innovation centre:

'The breadth of skills needed for these new future-generation products is very broad, and we found that we needed to house all these skills in the same physical location. All members of the team work along every step of the product lifecycle together, from ideation to prototyping to the manufacturing ramp using the same agile integration development methodology. Everyone is aware of the dependencies and linkages.' [10]

It is obvious that such an organizational set-up can only work once 'siloed' functional structures within a company are completely removed.

Another example is the digital services factory implemented by Schneider Electric – global specialist in energy management and automation. Schneider Electric wanted to reinvent the customer experience with new digital services. The company launched several initiatives focused on addressing challenges that would hamper the creation of smart connected products, such as long product development cycles, a lack of process methodology, and the duplication of efforts.[11]

To arrive at a viable digital services factory, the company decided to push ahead on two fronts: develop and integrate digital solutions with their existing products through an industrialized approach, and accelerate and scale new digital solutions. To complement both objectives, Schneider also wanted to gather data from connected assets across the company's infrastructure and customer sites to speed up the development of new services, from ideation to industrialization and market launch.

It built smart capabilities using a combination of real-time analytics, connected technologies and Internet of Things (IoT) platform solutions. At the centre of this architecture sits an IoT platform called EcoStruxure that connects Schneider Electric with devices from its range in use by customers. It leverages advancements in IoT, mobility, sensing, cloud, analytics and cybersecurity to deliver innovation at every level – from connected products to edge control to apps, analytics and services. EcoStruxure has been deployed in over 450,000 installations, with the support of 9,000 system integrators, connecting over 1 billion devices.[12]

The digital services factory eventually delivered on all fronts: incubating new ideas with a customer-centric focus, designing and testing potential product offerings, deploying and scaling offerings, and provision of analytics and IoT capabilities to accelerate application development. With the implementation of its digital services factory, Schneider Electric reduced creation and launch time for new digital services by 80 per cent. Its analytics-based insights help its teams to be more responsive and better anticipate customer needs. New smart digital services, such as predictive maintenance, asset monitoring, and energy optimization help their customer operations unit to be more proactive and efficient.[13] We have implemented similar digital service factory frameworks with close to 30 other clients in different sectors all around the world.

Marker 5: Setting up a digitally skilled organization to enable friction-free execution

A successful living product or living service company must combine the skills of what were historically three different companies: an Internet platform company, a software company and a traditional product company. Multidisciplinary teams with different working cultures must be woven together and supported by IT systems which facilitate this collaboration.

This vision is compelling, but the challenge is daunting as there are very real and very large barriers today between product development, manufacturing and service teams.

Bringing the best minds in the company out of their silos to work together is tough. It can create conflicts of priorities and resource allocation. And on the surface, this then often becomes a conflict between people At carmaker Tesla they have found a way through. According to a Tesla manager, with whom we had a conversation:

'Tesla's founder and CEO Elon Musk actively encourages engineer-to-engineer communication and strongly discourages hierarchical communication. A work chart at a more conventional company would show the primary communication occurring between managers who then relay the results to their engineers. Tesla profoundly discourages that pattern to the point where they quickly correct that behaviour in managers.'[14]

The company follows a very direct and lean 'least path' communication principle, which channels communication from the motor team directly to the firmware team and puts them together in the same room to work out the design. The Tesla manager explains:

'The technical architecture decisions and many of the core and even cross-functional decisions on how the product will be implemented are pushed

Figure 10.5 **Scaling up of new digital skills**

DIGITAL MINDSET & CULTURE		
DESIGN THINKING	• Maps the as is and to be desired customer experience • Creates prototype experiences for test and feedback • Designs application front-end flows
SCRUM MASTERS	• Communicate blockers (outside the team) • Remove internal blockers • Run daily stand up/demo/retro • Own sprint plan • Own scrum team metrics and release • Perform agile coaching
AI EXPERTS	• Have deep insights on all AI technologies from computer vision over machine learning to virtual assistants • Know which solutions help to increase efficiency of internal processes • Deploy AI in products and services to create new customer experiences
PRODUCT MANAGERS	• Own the product vision and initiate the product road map, articulating the epics and features that should seed the epic backlog • Create business case/s in relation to projects and/or backlog • Act as the 'customer proxy' ie the representative of the voice of the customer & colleague • Work with the other product manager to define and prioritize the stories in the program backlog
CLOUD ARCHITECTS	• Design and create distributed cloud infrastructure and services • Implement and maintain third-party cloud solutions • Ensure cyber security and compliance fulfilment across the entire cloud environment • Analyse cloud architectures to recommend ways to optimize performance and automate processes • Support data migration

GOVERNANCE
Strategic, operational, financial and customer-related decision rights

down to the engineering team. That secures integration among disciplines and engineering teams that is very hard to achieve if you are not putting people side by side in the same room, or virtual room, while they are working on the same problem.'[15]

Figure 10.5 gives a quick overview of which new digital skills need to be focused on.

Marker 6: Tracking results to constantly adjust course

The journey to the New is a long one where businesses are best advised to start with acquiring the seven core capabilities we listed in Chapter 9. When they then start to pivot to the New they must track their results and adjust where necessary. It is entirely a matter of good timing. Pivot too quickly or too slowly and you can suffer severe economic damage.

Manufacturers of smart connected products can experience true closed-loop, product lifecycle management where they track, manage and control product information and product functionality at any phase of the product's lifecycle around the globe. This creates a unique opportunity to continuously monitor and adjust the product and the business model in a way that was impossible less than a decade ago. Italian industrial equipment maker Biesse has, for example, connected thousands of its wood and stone processing machines around the world to work with the incoming data on new services and experiences.[16]

Leaders will embrace a process and culture that encourages rapid, iterative development and a 'fail fast and learn' mindset. Connected products can be constantly monitored and evaluated across this entire 'invent, incubate and industrialize' cycle, as shown in Figure 10.7.

Taking advantage of this data abundance, companies must develop key performance indicators (KPI) they can monitor regularly and feed into algorithms. Strategic, actionable insights can then be extracted to tweak the design and production of the product and improve the

Figure 10.6 From traditional products to smart connected products

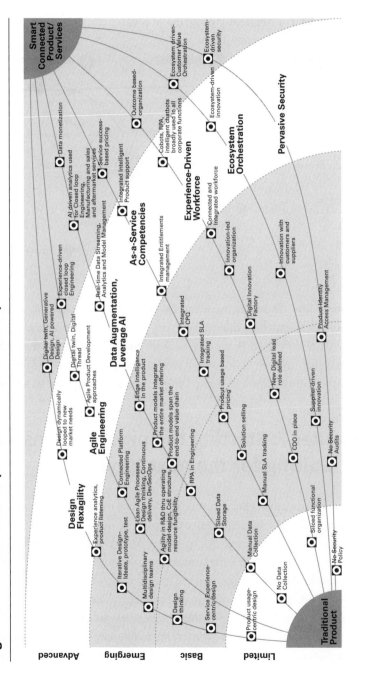

Figure 10.7   Iterative innovation process

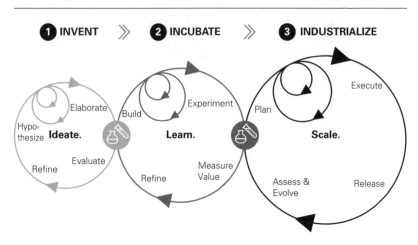

profitability of services associated with it. For this to happen, adequate digital infrastructure must be put in place to ensure that the insights derived from the KPIs are shared with teams relevant to the product's management. Even more importantly, the organizational DNA of the company must adjust to never-ending cycles of ideate, learn and scale.

Marker 7: Starting the pivot now instead of waiting for the next New

Digital disruption is here to stay. It threatens to displace around half the companies in the current S&P 500 over the next 10 years. If industrial incumbents can leverage AI to create smart connected products and complementary services that continually enhance customer experiences and establish them in the market at scale, they can dramatically boost top-line growth, and thus their market capitalization.[17]

Being too cautious in pivoting the business to smart connected products could be the cause of future failure. The markets for this new breed of product are expanding rapidly and will continue to do so – with or without you. Our advice is to act quickly and forcefully to ensure that history will remember your business as an industrial incumbent that successfully navigated the digital reinvention of its products and business model.

Takeaways

1 Virtually all product-making companies need to uplift the intelligence and experience quotients of their products, and to reinvent themselves into a living product or living service company.

2 To manage this transformation, a careful 'Rotation to the New' roadmap needs to be developed that includes digital transformation of today's core business to fund investment in the New. There are seven markers to help plot this roadmap.

3 Creation of a digital innovation factory is key for diffusing change and fuelling innovation as well as for attracting and retaining the skills needed.

4 Traditional organizational barriers need to be systematically torn down to encourage collaboration and increase agility.

11
Insights from the field

To help you enter the world of smart connected products, we have outlined the necessary core capabilities and defined a roadmap.

Nothing, however, is as compelling as reality. That's why this part of the book looks at the extent to which our analysis, theory and predictions resonate with, and are sometimes already reflected in current product-making practices across various sectors.

To get the layout of a business landscape in which many strategies are, as we speak, defined by smart connected products at various stages of maturity, we interviewed – or engaged in intense conversations with – a wide range of top executives, middle-ranked line managers and academics.

Among the testimonies are those of experts from software-making companies. In many respects, these are the role models for the more traditional product makers when it comes to developing smart connected products.

But the conversations 'from the field' also deliver valuable insights from incumbent hardware sectors such as industrial engineering, industrial equipment, automotive and consumer electronics.

What they all share is that, in one way or another, they have started the journey into the New. Their insights also prepare for Chapter 12, in which we will dive deeper into detailed case studies of companies in the midst of their journey to the successfully reinvented product.

Tesla

The trained software engineer we spoke with has dealt with smart connected products for most of his career. Having formerly worked for US mobility specialist Tesla, he shared his insights about disruption in the car sector and the great change demanded from traditional manufacturers if they are to master intelligent products.

'The same Tesla car a customer acquired in 2013 is today a much better car than it was when it was bought – due to permanent software updates.'

What is the key feature characterizing a Tesla car as a connected product?

I would say it is a car that has been uncompromisingly re-architected on a software basis. Tesla models are driven by software and not hardware architecture, despite the fact that there is a tremendous amount of high-end physical engineering in them. But it is software that controls the coupling of all those hardware components.

Can you think of any analogies in other product fields?

I would say that a Tesla model compares to a conventional car like a flip phone to a smartphone. For both very high-end electromechanical engineering is necessary. Both are very demanding to design and to make work reliably from a hardware perspective. But in one of them the user experience and the operation of the device is coded rigidly into the hardware – like in a flip phone. Tesla, rather like a smartphone, is by contrast a radically software-defined device that can change its function fluidly and very dramatically after the product has shipped. The beauty is that ongoing improvements of the user experience can be conducted via software updates.

What sort of improvements are we talking about?

This is a broad spectrum. We have to think of these vehicles as sensor platforms that experience and data-record the road life around them via cameras, ultrasonic sensors and radar. That data can be used to inform the development of driver assistant and autopilot capabilities or to develop driver-training sets. It can be used to compare the car's behaviour under human and automated control in order to tune the algorithms used in the autopilot functionality. The software-based existence of a Tesla gives you really a wide platform on which you can deploy and test any software that supports the car, from driving features to energy management to entertainment.

So, in a way, the car finds itself in permanent optimization mode?

This is crucial. Due to its quality as a smart and connected product, you can follow in a meticulous way how the product is being used. You see which features are more relevant to the customer and which are not. You can adjust interfaces along their actual usage patterns. You can prioritize features that are most commonly accessed and deprioritize those that are less used. You simply get tremendous product management insight from that operation data history. The product, being able to report its experience in the world, gives you a boost in developing features to make it behave in a way the operator expects and eventually appreciates the car as a good user experience.

Give us a real-life example of how software can change the user experience.

Take the first version of the high-acceleration mode in Tesla's Model S, for instance. It brought the car from 0 to 60 mph in 3.3 seconds. A simple remotely conducted software update helped to improve that to 3.1 seconds. It was a very significant improvement of the user experience and only doable with a connected and software-rich car. In more traditional cars the software would not be embedded deeply enough to influence such physical performance features. The same Tesla car a customer bought, say, in 2013 is today a much, much better car than it was when it was bought – due to permanent software updates.

How do you establish pricing models for fluid improvements of a car experience?

That is an interesting question. To pay an initial lump sum for a product and not much thereafter seems less compatible with a world in which you keep data connections with products in the field in order to keep them up to date and safe. As a manufacturer you have different expenditure for the staff and infrastructure to support the product over its whole lifetime. In that view the income is better stretched across the lifespan of the device as well. That is why making smart connected products drives businesses towards subscription models such as 'product as a service'. You have to decide how to decompose those services and how the initial cost for the mere hardware can be lowered. Attached to that is the important question of whether a product maker such as Tesla or another intermediary becomes the owner of the capital asset, being responsible for hardware maintenance. That would turn us into an almost Uber-style service provider.

In what way do you see the Tesla cars as product platforms having a function in a wider ecosystem?

Smart connected products all have a platform character in my view. In many cases they are not one single platform but a portfolio of platforms. Each of those capabilities is either provided by the product company itself or outsourced to somebody else. So if you build, for instance, a smart garden sprinkler, you have to ask yourself, do you want to develop your own technology with propriety sensors switching the device off when it rains or are you integrating third-party data from the national weather service to add that functionality? Our map data is, for example, taken from Google. So we are integrating third-party platforms into the product and it becomes then an interesting question of which of those platforms you insource into your own engineering and which you rely on third parties for.

Do these decisions not crucially hinge on how and where customer value can be translated into dollars within that ecosystem?

They do. You would, for example, not have thought that old-style taxi companies would come into conflict with automakers. But with a modern connected product that can happen absolutely. Who provides which service, is the question. And all of a sudden you have that reallocation of value amongst the different players in an ecosystem around a smart connected product. With that reallocation of value the core functions of a business can shift. In the case of cars, next to transport there is entertainment, there is delivery, there is payment and many functions more. All those functions have fundamental platform qualities that can end up as domain expertise for the provider of a smart product.

An engineering style such as Tesla's that is so vastly driven by data analytics needs a special company culture, I guess.

That is definitely the case. What I see unique with Tesla is that it is run with a very high appetite or tolerance for risk combined with an explicit approach to managing it. As a defining part of the corporate culture, this enables us to make substantially more rapid progress in product development and vehicle architecture than many competitors.

How is that reflected in the organizational set-up?

There is the universal recognition that all three engineering components that deliver the final product are equally important and need maximum dedication. There is the physical product itself, the combination of hardware and software that provides the operation of the vehicle. Then there is the development of the user interfaces integrating the vehicle's capabilities to a unique experience for the user. And then there is thirdly the back-end service infrastructure that monitors, manages and enhances the fleet of products in the field, enables improvements via software updates and operates the data collection. Tesla recognizes that all three things need to be executed extremely well; most companies don't yet have that done.

What about the expertise side of things; what kind of specialists work at Tesla?

Tesla is a Silicon Valley company. As such it is drawing on a pool of engineering experts who are very familiar with cloud applications. There are people who have done web application and mobile application development, for example in the infotainment industry. Only such a unique mix of software specialists can come up with a platform that can be merged with the best-of-breed automotive engineering. Most of the more conventional automakers are hardware companies without a pronounced software sensibility. Obviously that is changing. They are waking up and realizing that vehicles are on their way to becoming software platforms. But still they are by comparison primitive in the way they approach that. They tend to see the software as a lesser sibling to the hardware, as a kind of step-child of the company rather than a leading unit within the company.

How are information and data insights kept flowing within the organization so that they reach the right people?

Tesla's founder and CEO Elon Musk actively encourages engineer-to-engineer communication and strongly discourages hierarchical communication. A work chart of a more conventional company would show the primary communication occurring between managers who then relay the results to their engineers. Tesla profoundly discourages that pattern to a point where they quickly correct that behaviour in managers.

How is this concept put into reality in practical terms?

The company follows the 'least path' communication principle which channels communication from the motor team directly to the firmware team and puts them together into the same room to work out the design. The technical architecture decisions and many of the core and even cross-functional decisions on how the product will be implemented are pushed down into the engineering team. That secures integration among disciplines and engineering teams that is very hard to achieve if you are not putting people side by side in the same room, or virtual room, while they are working on the same problem.

Let's go back to the pros and cons of the smart connected products. Where do you see the biggest risks for manufacturers offering intelligent products?

One of the biggest risks is certainly security and regulatory compliance. Software is a field where you have to get something to work in a fundamentally hostile environment. That is much more difficult to achieve than in a friendly environment. You should put data security above everything to avoid damage done through attacks to the safety and privacy of customers. Malevolent hacks have become a not-too-uncommon occurrence in the world of connected products. Consider the recent botnet-distributed 'denial of service' attacks via IoT products.

The attack in 2016 that turned connected home cameras and printers into aggressive devices.

Think again what was happening there. A manufacturer of smart home devices wrote software, embedded it into products and then shipped those products out into the marketplace without the option to patch, update and improve them via remote software updates. What was effectively created was an army of remotely exploitable devices connected to the Internet. It was a bad service to customers. There is no denying that there are very substantial risks to companies when they do their software badly while digitally reinventing their products. It is just a high risk to not have the option to update the connected product's firmware, because it is just not possible to ship a product that is fully secure and connected. You do not know what attacks are going to come your way and you have to be in the position to mitigate those attacks through changes to the software.

Do you reckon something like that could happen to car manufacturers as well?

I have no doubt that it could. A similar thing happened back in 2015 when an activist group published an exploit of a model of the Jeep car brand that allowed the vehicle to be controlled remotely. Jeep had not prepared for this with adequate firmware updates, so they ended up having to ship millions of vulnerability patches (VPs) to their customers with new firmware for the vehicle. They had to ask their customers to go through a complex update procedure in order to mitigate an attack that was very much affecting the core safety of their vehicle. That kind of vulnerability is damaging to a business and the brand, as anyone can imagine. If you ship a product that is software driven you must keep the capability for security updates in the field without the customer being actively involved.

How disruption-proof is the car industry in general in your assessment as one of the main disruptors of the sector?

To my mind there are a number of incumbent carmakers out there who are the Nokias, Motorolas or Blackberrys of the auto world, who are not going to make the transition to digital successfully. It takes some time for that to happen but history in the auto industry will be written in the same way it has been written in the phone industry, I guess. Even the providers of the core product, a smart car, are running a risk of being commoditized. If you think of an autonomous vehicle then it is easily imaginable that such vehicles will become mere back-end infrastructure and the companies actually making money are the businesses providing the in-car services for entertainment, for example. These providers will buy their vehicles from a market of commoditized autonomous cars with not much technological difference between them and then monetize the transport experience and perhaps the advertisements shown when you have people sitting in a vehicle that is moving along the highway for a certain time. So you can see that huge opportunity for disruption of the sector. There is the old guard and the new guard. But even the new guard is at risk as the user experience in transportation and entertainment can change the economics and value distribution of transportation beyond recognition.

Automotive start-up in China

A smart connected product's success or failure hinges on the user experience. A former head of digital interface development at an automotive start-up explains what incumbent carmakers can learn from their new challengers.

'We worked like software developers – so that we could innovate and iterate the car experience quickly and very close to market demands.'

You designed the user experience and user interfaces of a new car model at a challenger automotive firm. Describe your principal approach.

We saw that the sector is about to go from a traditional manufacturing industry to a highly connected digital one where cars become like computers or smart robots. That led us to begin our work on a new car experience like a tech company, with a defined set of user needs or user problems to address.

What was the first step?

We identified all the details of core use cases, then drafted solutions around them. Only then did we use these thought-through blueprints to inform the hardware developers. This ensured that the features of the car would be a function of the user experience and interface.

That is a marked departure from traditional ways of conceptualizing and designing.

We saw the software industry working with this mindset for ages: 'Begin with the user'. It is natural for them because code is a living and breathing product that can be quickly iterated, which makes the process agile and flexible. There are standard procedures used by software developers, such as writing a user story, interviewing users, providing prototypes to be tested on their cellphones. Many of these methodologies are now bleeding into automotive, not least because new entrants show how effective that can be.

Has it become much easier to tailor a car experience in the digital age than in mechanically more rigid times?

The newer companies do indeed have some advantages here. User interfaces in cars are becoming more and more digital and, by this means, more versatile and targeted. The experience is increasingly based on software, which is highly adaptable raw material in a design process. I agree that this makes it much easier to put the primary focus on the experience. In our case it felt like creating an iPad for a car.

How would traditional car companies work on digital interfaces?

More traditional automotive folks still start a new car model with engineering requirement documents and integration plans, cataloguing thousands of specifications. They often still work with tier-one suppliers to make their user interfaces. This can make it more difficult to strike gold with a good user experience. We were able instead to feed our findings on future user needs squarely into our own agile development and design process. Most of the new challenger companies have all the necessary digital specialists in-house.

What was the information handed to your hardware providers?

They got from us a complete idea of what we thought users want from the product. We briefed them practically like makers of consumer electronics devices and not like vehicle manufacturers. We explained in great detail all the use cases we want to support, and what interfaces we therefore needed in the passenger cabin, including voice-activated, tangible and kinetic interfaces, what controls or displays were planned, and what the inputs and outputs would be. They also received our broader ideas for on-board entertainment, for passenger relaxation and how we think passengers can boost work productivity and avoid stress in their cars. One should keep in mind that when a car starts to move around autonomously, it turns into 'real estate' on wheels, multiplying the number of potential use cases. This dictates even more that the user experience should be the starting point and a major driver when vehicles are built.

Could incumbent automotive companies pull that off themselves?

Probably, yes. But there are considerable barriers for them in their accustomed competencies. Manufacturing, integration, safety, quality and testing have been the automakers' skill sets since they came into being. These competencies make complete sense. It is still much more complicated to bring a vehicle to market compared with bringing consumer electronics to market. If a home speaker does not work, it is a frustration. If a vehicle does not function it is a potential fatality. So the traditional car sector is built around reliability and quality. But that is also why the user experience is in many cases not at the forefront of developer minds in this sector.

Has this not changed as many carmakers have taken agility and digital technologies on board to make their cars more adaptable and user friendly?

To a certain extent. But again, it's new carmakers such as Tesla that are leading the way here. Autopilots and driving assistants are all software products. And they do not go through the lengthy testing rounds of a traditional car component. That makes things very quick. They introduce rapid iteration development even for hardware components, next to the software elements that are anyway pushed forward all the time for improvement of the customer experience. Because they produce software-rich connected cars, they can swiftly iterate on the user interface and driving behaviour of the car. However, in four-weekly updates there is more room for error than with updates every few years; this is a risk more traditional car manufacturers are often not willing to take yet.

Samsung

Yoon Lee, Senior Vice President and Division Head Content and Services, Product Innovation Head at Samsung Electronics America, analyses why large business organizations find it especially hard to self-inject innovation agility in order to master the world of intelligent products in the face of fickle consumer markets.

'Innovation is always a balancing act – there is a technology push and there is a consumer pull, there is Yin and Yang.'

Home appliances, your remit at Samsung, are becoming increasingly smart and connected. How does innovation work under these conditions?

Broadly speaking, refrigerators, washing machines and dishwashers innovate along lifestyle trends. But these consumer trends change at an increasing rate in the digital world, so that innovation processes have to pick up speed. Two traits prevail in today's home lifestyle. Number one: everyone is connected. Number two: the kitchen has turned into a hub. It has fundamentally changed from being a place for cooking into one for serving and finally into one for eating and living. Most of today's homework is done in the kitchen, and the most amount of family time is spent and communication exchanged there. We innovate products to remain compatible to those fast-moving market trends by intertwining physical and digital more and more.

That sounds like you only react to trends. Isn't innovation also about creating value for users that they have not discovered yet?

Innovation in our area is, realistically speaking, always a mix between the pushing of technological boundaries on our part and a permanent demand for novelties from the consumer side. You are right: what consumers can imagine is in many cases primarily encapsulated in their past experience. In other words, they do not yet know what they do not know. Picking up on Henry Ford's famous quote, consumers will say 'I want a faster horse' while the unspoken intention was to say 'I want to go faster from point A to point B'. Consumers will never be able to articulate that they want a car. The experience does not exist. It is our task to bridge that literal gap and proactively push for that extra bit of innovation that truly makes a product, the intended car and not just a fast horse.

How do you square the technologically possible with the genuinely desired by the consumer?

Innovation that is only driven by consumer pull tries to mend pain points identified via research or real-time data analytics gathered from connected appliances. It is a lot easier to do because once the pain points have been identified you can keep course as you aim at a clear innovation target. Pure technology pushes, on the other hand, tend to run into that initial reluctance hurdle, as consumers have no experience with the novelty yet. They tend to see the functional side at the beginning. Only gradually is the elegant experience discovered that comes with the innovation. We need the early adopters to kick-start a new product in the market. They care less about experience, whereas the mass-adopting customers of a later stage very much do so. Innovation in our field is always a balancing act – there is our push and there is their pull, there is Yin and there is Yang.

Does data analysis feature large in Samsung's innovation processes?

Consumer listening is an important part of innovating home appliances. We do have data analytics teams. We have teams doing core research. We have teams who improve data analytics algorithms. We employ specialists looking into the data to come up with usage insights. And we have regular meetings on how to improve products based on these findings. To provide AI cores to products and services is currently the biggest push of our innovation units. I see the main difficulty in the fact that consumer markets change all the time. The minute that you think that you figured it out and can march forward you put your head up and everything has changed again. I have worked for B2B manufacturers before and that happens less there, which makes innovation easier for those businesses.

How do you rate Samsung's overall innovation ability?

When I was a kid in the early 1980s I picked up a Sony Walkman and one from a Korean manufacturer. The difference was night and day. Forward to the mid-'90s, the Koreans had caught up. Businesses such as Samsung had by then gained core competencies mastering design, engineering and manufacturing technologies. As Samsung moved into the leadership position in consumer electronics, it was clear that the only remaining guiding light was the consumers. The product innovation teams were created in 2006 and the organization learned to properly read markets and consumer minds. It added getting something from research to market at relative speed to its core competencies. Then came the phase where 'rapid experience design' was introduced. With the arrival of rapid prototyping tools, both hardware and software, you would start with technologies first, to quickly build 'experience' for consumers to experience first and provide feedback to validate the unarticulated needs before proceeding with locking down the final product development specifications. It patched the problem that those who are doing consumer research have all technical capabilities to read what is going on in the air, but no technology competency to quickly translate those findings into tech solutions. So marketing and technology were eventually married very effectively.

Based on that, how would you define the current innovation phase?

All three stages I described were outward facing and developing our skill sets. The fourth innovation phase, today, is to put hardware, software and the entire business model into a new way of life within the company. This demanded that we completely change our culture of doing things in a holistic way from top to bottom.

It sounds as if that stage is the most ambitious of the three?

It means applying really revolving forces to the old ways of doing business throughout the organization. Innovation agility becomes the top agenda point. We need to be so quick as time has shrunk and clocks sped up. What once was a day of 24 hours has gone down to about three hours today. Human awareness, adoption, consumption and disposal of everything has become so fast that one day is worth three days in 1980s' calibration. That is why agility has become so important.

How can all watches be set to get to some coordinated agile movement?

It is interesting. There are people of all ages in the company, yet the decision makers tend to wear the 1990s' clock and the newcomers tend to be living in the 2018 time dimension. That provokes a cultural gap, but one that can be used to our advantage. For example, we try to mix maturity and experience levels within innovation teams, tearing down communication walls between marketers and product development engineers. That brings a lot of iterative culture into the company.

Does that really work? Does it stir up sufficient agility?

I think we get the central problem of many big organizations nailed by this. Think of the frescos painted in the Sistine Chapel. There are a few people working right at the ceiling and there are many standing on the ground looking up at the whole picture. We needed more minds working in a daring, agile single-mindedness on the changes right at the ceiling. That would make us quick. In an ideal world, we need almost to start painting first and do strategy later. The painters above are bound to make what looks like mistakes from ground level. The nose on a face might look too big, hands too small or the perspective all wrong. Having innovators do their work at such independent distance, reaping the rewards of being close to the market, can create friction with the finance and governance people standing on the ground looking up into the dome. By nature they don't like iterative or trial-and-error work or too much empowerment in the field. But I think that in the end we removed a lot of bottlenecks by introducing that culture; we just didn't have enough people working while hanging from a rope under the ceiling.

Samsung was relatively slow to tap innovative start-up impetus. At which point on the innovation learning curve you describe has that changed?

It is true that the organization was not shelling out billions right from the start to buy start-ups. In fact, bigger acquisitions started only from around 2010. But Samsung has caught up here as well with quite sizeable and bolt-on transactions. A designated organization has been created to secure a maximum of technology transfer and suitable hirings from founding firms. But by being slower than others, the business could perhaps learn more about how to make the most of these acquisitions.

In what sense do you mean that?

Companies get big when they scale their business model. But as you grow, inevitably the inertia and complexity in your company multiply. As a remedy, people start to indifferently prescribe start-up angles within the mothership, which is effectively a dose of chaos. This typically doesn't work unless there is a role within the mothership that understands both the start-up culture and the mechanism of the mothership that can effectively guide the agility into the mothership's culture and processes. In large businesses it is better to create smaller organizations that start from scratch. But start-up organizations, self-created or acquired, cannot scale, they are just movement and results. Their mothership organizations, on the other hand, while they may find it difficult to be nimble in creating new value, are good at moving in step along clear processes in scale. They are fast if they have clear goals and directions. So, in practical terms, identify first what part of your business is supposed to be gaining more agility. Then obtain the necessary capabilities fast through acquisition or by internally designating a small team with independence to develop suitable concepts with agility. When the concepts are successfully built, mature the concept by building clear directions to fit better for the mothership organization to execute at scale. This 'tailoring of agility' should be led by the internal team to the mothership team. This team's role is critical to the success of scaling the new agility. Big companies need to build this type of organization within the mothership organization that can act as the conduit in guiding agility into the mothership, like a tug-boat. That is what Samsung did on many occasions. That is what creates agility in large organizations – not necessarily mere acquisitions.

Dassault Systèmes

Olivier Ribet, Executive Vice President, Europe Middle East Africa Russia at French software maker Dassault Systèmes, believes that only the extreme blurring between smart products in the field and their digital representation at their makers can secure a lasting and adaptable product experience.

'When you connect virtual to real, people imagine, invent, build, manufacture, collaborate, distribute, sell, maintain, repair and service all in the same environment.'

What is the essential rationale for using representational technologies such as IoT software and 'digital twins' in the manufacturing of smart products?

A lot of people speak in the context of intelligent products these days about product-as-a-service approaches, about not owning products anymore but only using them, about pay-as-you-go billing models. However, a smart connected product is more than a device plus an antenna. Unless you engineer the product's complex user experience from the beginning, it is going to become extremely difficult downstream to reconnect it to be truly connected. That is the essential logic behind these technologies. From the first day you think about a smart connected product, you don't want to separate development and engineering from product usage and ultimately the end-to-end experience it provides.

That sounds, at first hearing, like a tall order.

It might, but I can tell you that only when you closely and realistically connect the virtual to the real life of such a product, will people imagine, invent, build, manufacture, collaborate, distribute, sell, maintain, repair and service all in the same environment. This is absolutely necessary given the complexity of intelligent products. All these teams have to know about or actively influence the twin data and, by this means, the product itself in iterating editions. In other words, if you cannot compare the data from the real world with a true digital twin representation and thus with the product experience you want to provide, you cannot take the necessary decisions.

Give an example of how efficient such technologies can be for product makers.

Let's think of a smart washing machine. A decision is taken to enter the Japanese market with the product. This machine has to be small, portable, easy to install in any city in Japan. The concept also says it has to be able to wash more than 200 times per year and it cannot cost more than €200. It has also been decided that the device has to offer the functions of washing, drying and, as an additional feature, humidifying the room. Plus, this machine cannot weigh more than 150 kilos, and has to withstand 200 Newtons of impact when it drops from a truck. All these features and functions – and in real development processes of course many more physical properties and logic functionalities have to be determined – can be represented in a 3D digital twin available to anyone having to deal with modelling, simulating, producing, marketing or servicing the product.

What is the big advantage for all these people and teams involved?

Once you have decided on these parameters, you realize that this is both a business case and an engineering question. The product managers and marketing managers are jointly building the product. This is an IoT device, so all these managers have to come together. This is a product that will regularly report back to them what it is doing, what its performance and usage is and what maintenance it needs. On top of that, this smart machine can adapt itself to use cases such as a 70-year-old lady living alone or a young family with two small children. This dream of a highly adaptable and responsive device can only be fulfilled when, from the beginning, all systems and subsystems, all functions, process logic, and physical constraints are being digitally modelled in tandem to be thought through, simulated and tested virtually – before any real version of the washing machine comes off any assembly line.

How would virtual simulation, testing and optimization work in this case?

With the digital twin you would simulate every aspect and episode of real-life behaviour, down to little details. So, you virtually press the start button, you virtually put in water of a temperature of 75° Celsius and pH of 2.6 and then you watch what happens. Everyone involved can see digital continuity from design to mechatronics to software system. All the disciplines are managed and governed together in one data model. So, it is not just some vague marketing dream. It is a perfect representation of all engineering aspects of the original that is not real yet, but can be. All components have been selected and you know what you will build as you have a perfect facsimile of what the people on the production lines will be assembling. At this stage you could, for example, bring in local electronic retailers and discuss with them whether they would put this machine with these defined specifications into their catalogue.

In what way does the digital twin remain involved once the machine has ended up in, say, a real household in Japan?

When the product leaves the factory, the real life of the product really starts. At that point you as the manufacturer of this smart product start to learn. Your product phones home saying, 'I was designed to have the water at 75° Celsius at pH of 2.5. But I actually run at 62° Celsius and pH of 5.6 here in Kyoto.' That data arrives at home and feeds right into the digital twin. It can then be used to simulate the situation and send data such as alerts to the user or operating software. Changes can be made and conveyed back to the machine in Japan. The hyper-realistic twin allows for the manufacturers to move from modelling to making to learning to simulation to continuous adaptation of the product – all in the name of a good user experience. When our customers want to create a smart connected device, a robot or a car, say, they think about user experiences in three dimensions: connected experiences, contextual experiences and continuous experiences. That is exactly what a virtual 3D twin should allow us to simulate and optimize.

PTC

James E Heppelmann is President and CEO of PTC, a maker of IoT platform software critical for the development of smart connected products. In his daily client dealings, he realized that most industrial companies face significant challenges and that often a root-and-branch recoding is needed when it comes to entering the world of intelligent products.

'You don't go to bed as an industrial company and wake up as a software company – it is a lot harder than that.'

It is argued that software companies find it much easier than their hardware counterparts to adopt the know-how and processes needed to manage smart connected products successfully. What is your take on that question?

I would say that it is definitely more natural for software companies and I actually think the whole concept of software is still a bit of a foreign topic for many industrial companies. Even very big industrial businesses have declared themselves software companies and have then realized that it is a lot harder than that. You don't go to bed as an industrial company and wake up as a software company, after all. It is more complicated than that and it puts that whole topic of cultural transformation on the agenda.

What kind of cultural transformation is needed for them to become successful in these new business lines?

Things are getting more complex and speedier and that has to be dealt with. Just think of the clock cycles of new digital products. We as a software maker can, in a natural process, create a new product weekly, even daily if we wanted to. But a physical product can be changed only periodically and that can be a very expensive process in terms of new tooling and all the other things necessary in the hardware world. Given software companies are used to the fast iterations of software development, they find it a lot easier than industrial businesses to keep a fast innovation pace. A company that makes smart physical products essentially has to be a hardware and software company at the same time. You are not simply switching between these options. You are trying to do both simultaneously. But these two strands involve very different processes and very different cultures. Think of one company trying to act like two. Not as a merger but actually two types of businesses in parallel and that is very challenging.

What can they do to keep their businesses firing at full power on these two strands?

In the end we all have to take on board that software is a key part of the world going forward, but also that the physical element of products is not likely to go away, but its contribution to total value is likely to decrease. And what is for sure is that you cannot ignore the software components altogether because you then get squeezed out of the market if you don't have that new value source. So, I think companies have to figure out their individually balanced operating model, having some people doing hardware but being mindful of how software come into the game and some people doing software whilst being mindful about the hardware parts. To keep these two realms aligned even while you have people doing separate things within them seems to be the solution to the riddle. But that is not easily done.

Next to cultural challenges and adopting new mindsets, what is the highest practical hurdle to overcome, would you say?

The most important one is getting enough speed. Software can so quickly be changed and pushed into the market via downloads. The same pace is hard to achieve on the hardware side, where the unification of development activities with operational feedback from product or user is more difficult to achieve. How do you monitor what is going on and how do you let product changes happen quickly without breaking anything? I think that is a very big challenge for people.

What about the most urgent organizational adaptations to be made?

It is certainly important to find the roles of actual engineering versus IT in developing smart products. That is a central but still widely unresolved question. Because frankly the traditional engineering teams don't know that much about the data concerns, the security concerns, the failover concerns, the cloud analytics capability of the business. That is a whole new world to teams who work on physical products. The IT people within the organization know a lot about that. But they have never been in a product delivery process. They are typically running internal systems such as CRM, ERP or PLM. And now they have to all of a sudden contribute to a delivery process. So again, things can get complicated here. You have two organizations that in a lot of companies frankly do not like each other that much. And you have to bring both of their skill sets together to create some new model that marries the best of what they each know in order to get an 'evergreen' product support model going. To my mind only a very few have found the right formula so far.

Can perhaps small and agile business organizations such as start-ups bring skill sets on board to get that amalgamation done quicker? What can your experience with clients tell us here?

I don't think that the start-ups are the answer. Because sooner or later the main organization has to have the full skill set operational. I suppose you can work with a start-up but only temporarily to develop your own freestanding capability throughout. The difficult stage will be again to transfer and scale that capacity so that it has a business-wide effect. If I look at my customers, they tend to create their own new digital branches internally that look into smart products and then maybe they start to co-opt people from IT into it. They would rather divide the IT team into one that works on the product development side and one that continues running the business systems. I think that is pragmatic and works in many situations.

Entering smart product lines means, as you have pointed out, drastic cultural and organizational change but also substantial financial commitment. Do your customers find it easy to get to the right formula when they invest into skills and enabling software such as yours?

I have been selling software for much of my career and I find customers very motivated in the dynamic phase we are currently in. The world is changing so fast and a lot of these product companies are afraid of becoming dinosaurs. They just fear getting squeezed out of the market. And they may not know how to immediately react but, from what I see, for that reason they are even more motivated to explore new things. If I compare it to earlier phases when it was about adopting more conventional CAD or PLM software there is now much more preparedness to try to figure out how IoT software can put them on a really new trajectory.

How much self-confidence is there to really pivot into the digital New?

It is still a hard job for customers to figure it out and it is certainly a big concern for them how they can monetize such a big investment. But these considerations are counterbalanced by the question of what competitors do that they don't do and how their sales can be kept going in the future. But even if a management team is not tech savvy they are feeling the pressure that digital transformation can do great things for them. At some point in this limbo they have to turn to somebody on the technology front and ask: is this really true? Can I actually believe this stuff? Because I am struggling to get to an independent judgement.

Who should in your opinion find the answer to that question; who should lead the way into digital in a traditional business with not much digital experience?

I think if you have a visionary CEO that is great. But let us be frank. Many businesses don't have such a figure at the helm. Many have more financially minded or sales- or marketing-minded CEOs. So I think you do need a leader in the company who champions the cause. A CEO can step up and be that figure. If so, that is great. But honestly a lot of CEOs are afraid of this topic. For someone who grew up surrounded by physical products all these new software implications can be a scary place. So you need to find that fearless leader type to overcome all the objections and the more highly ranked that person is the better. I think, by the way, that it is worth finding such a figure at executive board level as well as at supervisory board level. And it should be an urgent impulse for supervisory board members or non-executive directors to robustly challenge their executive board about their digital strategy.

Can you think of a real-world person that has had this impact on a large business organization?

I think, when Jim Hagemann Snabe, a former SAP executive, joined the board of Maersk, the Nordic shipping business, he started a huge fire about getting digital. This appointment seems to have changed many critical things at this sprawling company and it underscores the point that in most cases you need heavy-hitting digital evangelists coming on board to drive that necessary change process towards digital in complex business organizations.

Coming back to PTC's software platform: how do you conceptualize a product such as ThingWorx to make it practicable in corporate organizations where experts from all walks of business life must deal with it and cooperate through it?

The main thing we try to do with our IoT platform is to make it easy to handle for a 'citizen developer'. By that I mean somebody who is computer savvy but not necessarily a software developer, somebody who is much closer to the actual business case of the company. That is why, for example, we use a lot of drag and drop features with the platform. Because in support, in manufacturing, in development, in sales, in all these functions you find people who want to take that data coming in from smart products and put it to work quickly. Each of these groups needs tools that allow them to configure an application or a point of view based on that data that would help them sell, service or manufacture differently and in a better way. So you find hundreds of creative ways to use the data transmitted by smart connected products but it will not work if all that splits into different software development projects. There is just no time for that. Hence you need a platform that makes it quick and easy to build a new application for all these occasions and audiences. That is where we see the value of our software to get to these applications fast and change them over time with minimum hassle.

Your IoT software is itself designed like a smart connected product. What does it report back to you and where does this help your and your clients' businesses?

Our software is used in so many of our customers' business functions that we really have to be creative to figure out the business cases that create value and therefore need our focus. We know across our customer base how the different versions of software are being used and what the distribution of use is. We can also use that data for sales analysis. We see if customers are successful with their deployment, whether we should send somebody from our customer success group to give support. If they have a certain number of licences and we see that they have routinely hit the upper limit we should get in there and sell them more because they obviously need more. So there is for us a technical support use case, an engineering use case, a sales use case and a customer success use case.

In what way can you use the data to improve your product?

We know about product failures such as bugs for example. A new release gets out and we may have dozens of bugs reported to us. We know through the closed data loop with clients that some of them have occurred only once or twice but that others have occurred perhaps thousands of times. So we can take a decision to fix the bugs where we are getting the most 'bang for the buck' quickest. If you will, the capacity to be so closely connected to our clients gives us the opportunity to pick the lowest and at the same time biggest fruit. And that is of big value to our own business model while it helps our clients.

Could industrial companies not develop such software platforms themselves?

Manufacturers might well have originally thought to develop such technology. But I think given the speed at which everything is moving, and all the new use cases that are coming out, people realize that it would be a tremendous effort to do it with only one end customer benefitting from it. We also think that an IoT platform for the smart connected product is the classic case of the 'long tail' problem. Twitter is one application used by millions of people. If somebody is trying to do smart connected products they start with one application to monitor the fleet. Pretty soon they need another application to try to do analytics and predict downtimes and then they might need another application that will inform sales and account managers and yet another application might inform them what might happen on the customer side. So we talk about a whole string of applications that might be used by very narrow audiences but would create tremendous value. A universally adaptable platform is the right answer to that in my view. And industrial clients seem to have taken that on board – after all we are not the only vendor in this IoT platform market.

To what extent have you turned into your clients' consultant on smart connected product use cases or even business cases?

We stop short of giving specific product design recommendations. We do not say to our customers, say, these six points are the most promising feature set for your product to work in the market and create value.

But we give a framework for how to think about it. We tell them broad options: you can try to differentiate your product and make it 'evergreen', you can try to change your business model, you can try to add value through more efficient operations. We are a technology company and not a consultancy per se. We consult as far as it is needed to make our technology work and look for consultancy partners to take it from there. We don't have the capacity and are not smart enough about all possible applications around our technology. So we like to stay on the software side creating enabling technologies.

Caterpillar

A retired product director responsible for product development and innovation at equipment engineering giant Caterpillar explains how the construction sector can dig for lucrative value deposits with the help of digitally orchestrated machines.

'Roads tend to be 3D printed these days – and only smart connected machinery can pull that off.'

What drives the trend towards smart connected machines in the construction industry?

The drivers are mainly cost efficiency and the demand for higher precision in project execution. The road construction sector, in particular, has a lot to catch up with here. Construction sites can be run today with much higher operational accuracy and cost efficiency, due to the technology that has become available.

Let's dissect the cost argument first. How do your customers look at this?

Construction up to now involved a great deal of human planning and uncoordinated action that left teams and machinery idle for long time spans. This means that, for example, orchestrating the movements of a truck fleet around a road construction site by digital technology offers a lot of efficiency gains – cost-wise and also operationally. That central nexus is recognized by more and more customers who become willing to work at higher precision levels, for instance by optimizing the number and kind of machines on a site depending on daily changing parameters such as specific ground or weather conditions.

What is behind the demand for higher building precision?

More precision feeds into cost effectiveness as just described. But building a road also works today in many cases similarly to 3D printing. A new road needs to have precise curvatures and drainage slopes. More and more of these construction sites are now being 3D-designed at planning offices and machines are expected to download these digital blueprints and give them shape in reality – just like a 3D printer is expected to print mechanical parts. The commissioners of road projects such as ministries and other public bodies are also requesting more precision. They have become noticeably more cost sensitive and now issue much more detailed tenders based on very different charging models compared to the past.

Describe today's approach compared to the past.

Commissioning parties increasingly want to be charged on the basis of 'kilometres of road built' or 'cubic metres of dirt moved'. And they often want to be invoiced for accomplished tasks on individual days. That is quite a departure from the past practice. In earlier days contractors would have cost-quoted a whole road project including a financial safety margin in case of unforeseeable extra costs. Instead, today's commissioning practices drive contractors to get their estimate as close as possible to the real cost. But this works only when they can be sure about precise performance levels of their machinery – and these levels can only be reliably achieved if digital technology orchestrates the machines. As a bottom line, construction today must have a much better overview on what they have achieved in what time (real-time job site management), how many cubic metres have been moved and so forth. Otherwise they risk a project becoming unprofitable.

Give an example of how this high-precision machine set-up can look.

Let's build an imaginary road. At the centre of such a project works a paving machine laying out the asphalt. Behind this machine, up to four compactors are following, giving the asphalt its final density. High-quality construction materials such as asphalt have become so demanding that the first compaction has to happen within a certain temperature range while the second one needs a different temperature range. Start too soon and you will try to compact a cream pie, too late and you cannot compact at all as the material has solidified. For this not to happen, the paver transmits to the adjacent compactors the temperature of the material being laid out and information on what time windows the compactors now have to do their work – depending on weather conditions (air temperature, wind, sun and ground temperature) on the day. You can see here that advanced road construction today better takes place within an ecosystem of machines securing high quality standards as unintended misuse of material and machinery can be reduced.

How big can these machine ecosystems grow on project sites?

The bottleneck is always the paver. It can lay out up to 350 tons of asphalt per hour which is also roughly the capacity of a standard asphalt plant. This means that up to 60 trucks have to be lined up in front of the paver to deliver their hot asphalt loads (depending on distance between paver and asphalt plant), one by one right on time (asphalt can solidify in trucks if wait time is too long). This can again only be done if trucks are connected and managed by a supply chain system connecting the paver with the truck fleet. The data loops are far-reaching. If the paver capacity is for some reason momentarily reduced the asphalt plant can slow down production because it has also been sent a data signal. The paver can also be made aware of the location and traffic situation around the trucks and the asphalt plant is aware of what happens on the site, while the paver also manages the compactor fleet. You can see how this kind of interconnected intelligent working style can kill a lot of inefficiencies built into old-style process chains in road building.

In the clockwork you describe there seems to be hardly room left for individual human guesswork, decisions or personal experience.

That is right and that is how it is supposed to be. In fact, we are aiming at squeezing out human decisions and actions altogether, saving supervising roles on site. At the current stage machines have taken over the capacity steering and timing roles, as I described. Human operators still trigger most of the actions – for instance on a paver or compactor.

This will eventually be a thing of the past?

Yes, over time. Today a paver still needs three full-time operators on board, but the machine already carries 12 computers. On-board intelligence is growing so that the machine can eventually become autonomous. We are also working on making the compactors unmanned as their precision work is so important for the final quality of the road. The machines will then be connected to an intelligent expert system orchestrating all the autonomous machines on a project. This is a set-up where a maximum of cost efficiency will be reached. We often compare the construction sector to the agricultural sector where a similar cost-driven trend towards semi-autonomous and eventually autonomous machines is underway. The difference is that they work in 2D whereas we have to master 3D and have to observe more safety aspects as roads are often built in densely populated inner-city areas.

Who controls the complex data flows in autonomous road-building projects?

Our rule is: the customer is the master of all the data that is created. So, we see ourselves as mere architects of such orchestrating expert systems and as the neutral operational monitors of real projects. We would create the apps operators use on the machines, we would create the backbone IT systems, the telematics units and we would make sure the data loops transmit the right data to the right party.

But you are not part of the real-time data communication itself?

Yes we are, if not at operational levels because this is the remit of the customer. But we have a lot of data running through the system that allows for predictive maintenance and monitoring. We gauge cooling temperatures, oil pressures, downtime intervals and all sorts of other data the machines collect via their sensors and transmit. And we communicate back to the operators on site when we see something that is not working or hampering the workflow. We employ designated teams for this. Each team member monitors around 200 connected machines at a time and picks up warnings by the expert systems when something does not work correctly. The only aim for this is to keep safety levels as well as uptime of all machines as high as possible.

Do all these data insights also feed into your own innovation process?

From earlier, less connected machinery we already knew a lot about downtime patterns and how machines would behave in the field. But that was based on empirical data taken from unconnected 'black boxes' attached to the machines. In contrast, we now have real and well-structured operational data available in real time, which allows for much more precision. We can actually see what a connected digger or truck is used for, what components of the machine come under which kind of mechanical stress on which application. That gives us a lot of insights for product adaptation. You can now correlate, for instance, the temperature data of hydraulic fluid with the performance data of the hydraulic pump and extrapolate from that the lifespan of a component. So, the data received from smart connected machines feeds right into product development.

Do refined data findings also segment different markets and can you tailor your products now in a better way?

Well, we learned, for example, that the load cycles on construction sites in China are much quicker than in Japan where idle periods last surprisingly long. This might have its cause in higher congestion on the roads in Japan, slowing down trucks. In any case this has led our engineers to think about whether construction machines for the Japanese market should not be optimized primarily for fuel consumption than for performance. On the back of these findings in one vehicle model some parts could be downsized and new vehicle lines created for different markets launched. That gives you a competitive edge and it is again a direct translation from a very pointed data finding collected from connected machinery.

You call the various connected machines an ecosystem. To what extent do these connected machines function as open platforms?

We keep the software system closed. In that regard our connected machinery is not a platform concept. We are not creating a Linux solution open to third parties when we create an expert system for a building project. The main reason is operational safety on the site, but another reason is data safety, as attempts have already been made to hack into construction equipment. In that light it is better to keep our system closed and thoroughly encrypted.

When did the shift towards connected equipment take place within Caterpillar?

We started out 25 years ago with the first solutions for product health monitoring which involved mainly maintenance optimization. Machine hours, machine location and a few more data points were collected and analysed for insights. Little by little that picked up and the organization eventually realized that if we stuck to being predominantly a hardware provider we were in danger of becoming a commodity and our customers were expecting much more from us. We realized that we had to be more than a hardware producer and then, about 10 years ago, we firmly understood that we also had to become an intelligence provider or we would lose market clout to disruptors.

What were the concrete steps taken to achieve this?

We hired a bunch of data specialists who had no background in industrial engineering but focused on statistics and big data analytics. We created university links and opened a lab where new data mining technologies and expert systems are brought to maturity. And we also brought in numerous new, mostly younger managers to make sure they were focused on the software side of things and not only on the hardware aspect. All of that was then put under the umbrella of a new division in our company called CAT Electronics. So, it was all in all a mix of different initiatives that helped to build the new organization.

I believe that thanks to its unique position of world leader in the industry, with its technology know-how and comprehensive product line, Caterpillar is in the best position to offer intelligent ecosystems of machines to customers; they can continue to be cost efficient on job sites and win more deals with our totally integrated solutions.

HP Inc

Bill Avey, Global Head of Personal Systems Services at global tech company HP Inc says that over the decades many hardware competitors have fallen by the wayside. In that light, he explains, companies that are able to combine great innovation with a product-as-a-service strategy will stand out in the marketplace and lead the industry forward.

'Hey HP, we really like what you do in the print arena – could you also do for us in the PC space what you do in your managed print services?'

What gave HP the impetus to embark on its distinct DaaS strategy?

Next year will be the 80th anniversary of HP. There are not many hardware competitors left who have managed to navigate that highly competitive technology space for so long, and that can only be done by constant transformation. 'Device as a Service' is something we have been doing in the managed print space for business customers for a long time, because we saw this as a decisive competitive edge in a maturing market. That is the main part of the genesis of this model at HP. We eventually extended that model into the consumer print space with schemes like Instant Ink and a subscription model that has been received very well. And we then extended it to our PCs as well. We are currently building a 3D print business right now and from the very beginning it is being created as a DaaS model to give us the edge in the market. Right now, all across our business, we are finding ways to deliver industry-leading HP innovation as a service. This model plays an important role in our future strategy.

When did HP turn that into a declared company-wide strategy?

About three years ago we were getting more and more demands from our customers saying, 'Hey HP, we really like what you do for us in the managed print arena. We also like your PCs and the way they are designed to go with services. Could you also do in the PC space what you do in managed print services?' They were basically looking for a partner that can provide devices in combination with all sorts of continuous services – beyond the bundle of services and accessories we were offering them anyway at their discreet demand. Based on this new demand, we started what we eventually called 'Device as a Service' in 2016. These comprehensive device management services enable customers to modernize legacy IT environments in a smart, efficient manner. By having HP manage device fleets as a service throughout the entire lifecycle of the technology, IT departments free up valuable time and resources to invest in strategic growth initiatives inside their own organization. It's a win-win.

What role does the smart connected device play in that strategy?

It is a very central one. We knew at the time of its introduction that the DaaS model would only be brought about effectively through smart connectivity. So our devices, for instance the PC range, are IoT items that can tell us how the device is performing and how safe it is. We created software for remote analytics and proactive management of those device fleets. The software collects and analyses hardware performance data that can be used to enhance the user experience on each individual machine.

Expand a little on how that works in practice.

For example, when you get to that moment that one of your PC's batteries needs replacing we would be able to anticipate that before it happened and proactively ship you a battery. Or, say there is an executive assistant who worked in a sales department for which we had initially provided the right device. Then this assistant receives a development opportunity and moves into a marketing department. All of a sudden application software such as Photoshop and Auto CAD becomes part of this person's day with the need for more storage and processing demand on their device. Via our analytics tool we can set the alarm bells to give people what they need at any time. We can say, hey this person had the right device over the past 15 months but three months ago it all changed. The machine is telling us that there is something wrong and we can now say why and go back and actively remediate it.

Where is the cost-saving effect for your clients here?

There are many that we can provide through the DaaS model. We would, for example, make sure that your IT department runs a virus scan and test your firewall frequently which helps in keeping costly data security problems outside the company. But we can also help our customers understand whether they have underutilized or overutilized an individual PC or processor. We can help them understand whether or not the apps installed on the machine are actually being used. This can help a customer to find out if they are overlicensed or underlicensed. All that insight can drive savings for the customer.

Is your DaaS analytics software working across various operating systems and hardware vendors?

It must be able to straddle a wide portfolio. Many of our customers are based on Windows but there are also creative departments and these guys often go for Macs. There are very few clients who have a pure HP environment. Over the years they might have bought some of our competitors' devices and there are very few among them who have a pure Windows environment. From day one we therefore wrote that remote analytics software for a multiplatform world. It bridges an environment where, for example, Dell, HP and Lenovo devices work in parallel. Essentially any Windows-based device can be connected to that system. We also wrote for a multi-OS world so that we could deal at the same time with Windows 7 or Windows 10 as well as iOS and Android and just recently we added Mac OS. This broad approach has had a lot of resonance with our customers.

You as a vendor have powerful relationships with big device buyers. Can you use that power to attract other device makers to become partners in your DaaS scheme?

The fact you describe actually led to the discussion with Apple and the integration of their operating systems and hardware. I cannot speak for them, but they had a great interest in being part of the offering. Understandably they get their devices into the enterprise world more rigorously than without being part of such an alliance. Exactly for that reason these devices would be actively managed by our software. Now our customers, through that compatibility, can have their entire Apple portfolio managed by HP; that includes the desktop devices, the notebook devices and the mobile devices including items like the watch.

You mentioned firewalls and anti-virus scans. How important is data security to customers in the context of your DaaS offering?

The market has certainly evolved way beyond procurement of devices and services. It has moved deeply into security. That is probably one of the biggest differentiators of HP. We pride ourselves on making the most secure devices and also on keeping them the most secure through our management solutions. People have very often failed to recognize that, for example, printers are IT endpoints in commercial environments with memory, with storage, with an operating system. When you have all that, you have to be protecting against security threats. Too many people do not realize that they have vulnerability on that end. That is where a lot of differentiation comes now and where HP is focusing to create the most secure solution. Our competitors do not move a great deal into other tech areas. We compete against them and have to make sure we have an exclusive competitive value proposition compared to them. And security is certainly one of the issues. After data breaches have happened, which can never be ruled out, our analytics and proactive management software helps customers to get up and running again quickly and to limit the impact of a breach.

And you have gone into relationships with external partners in that field as well. Why?

Our DaaS model also foresees the cooperation with providers of cyber-security insurance policies – exactly because security has become such a big topic with our clients. We have gone into a partnership with Aon for a cyber-security audit that is provided to customers. Based on that audit, Aon issues a cyber-security assessment on how vulnerabilities should be addressed with a client. Often one of their recommendations is to use HP devices or services involving analytics and proactive management to remove such vulnerabilities. We are not paying them anything for that; the reason they recommend us is that our managed services make sure that anti-virus screenings are done properly and firewalls are switched on effectively. By recommending us they have the security that the protection levels are implemented and maintained. And in a further step, Aon offers favoured terms on its cyber security to our customers – an advantageous partnership for all three of us.

Lastly, describe the typical customer of your DaaS offering for PCs and what pricing model is typically picked.

I would say that the average services customer is one that is procuring the devices from HP, combining that with a standard bunch of device life-cycle services to have it up and running. These services would comprise, next to the managed analytics services I talked about, Windows imaging, managing the right BIOS settings, and physical as well as electronic asset tags. It would also involve the actual deployment of the device, the data transfer from old to new device, expert support as well as asset recovery of the old machine. All of that would be bundled up into a per-seat-per-month fee. This fee is made up of the services, the device and the term, which for most customers is three years. But these parameters vary. For one customer, a global restaurant chain, we have a per-device-per- month model for the cash registers. Or take our ink subscription programme for home users. It basically allows our consumer home office folks to sub-scribe on a monthly basis to a certain number of pages printed.

Mindtribe

Twenty years ago, Steve Myers founded Mindtribe as a start-up to seek a better way to build innovative hardware products. Since then, as a CEO he's led the creation of some of the world's most successful hardware development projects at large companies and start-ups alike. In doing so, he pioneered a distinct agile product innovation formula that guides client decision making and elevates product quality, business impact, and overall user experience.

'There is no limit to the trend towards connected products, it will become the fabric of our existence.'

You lead Mindtribe's 40-person engineering team. Your firm combines a rare depth of technical expertise, creativity, and product management experience. What new categories or new types of devices is the developer community currently excited about?

Within these communities there is a wide range of products with connected capabilities in development at a given time, but classes of products often come in waves as markets evolve alongside the emergence of higher-performing, lower-cost technologies. For example, around the year 2014, we saw a first wave of connected products with a focus on consumer wearables such as fitness trackers. Now, in the next evolution phase, I see a lot of innovation happening in devices around the transportation and health sectors, for example mobility concepts for all types of ways to get around in a city – from smart bicycles to scooters to motorbikes and shared rides in cars. Innovators will not just be focused on technologies for autonomous cars in this field. Think, for instance, of ride-sharing companies like Uber branching out into e-bikes and scooters in various cities in the United States. In health, I see a wide range of new products at the intersection of consumer products and traditional medical domains, such as devices that allow consumers to both monitor and manage their own health.

Getting smart products to market at speed is not easy, especially if the technology stack bringing them to life gets more and more complex. What are the troubles product makers are facing here?

Connected hardware development is somehow broken today and hasn't benefitted from a lot of thought leadership. Way too many hardware development efforts result in products that aren't very good, never ship, or take too long to develop (which also costs too much). In a nutshell, there is a lot of wasted time, effort and money in developing hardware products. The number one problem is hardware teams simply making the wrong product. Hardware teams usually work in too much isolation from marketing and product and business strategy. Because hardware development and iteration timescales are relatively long, say compared to software, that isolation means teams can be making the wrong thing for a long time before they realize it, after investing a lot of time, effort and money. These epiphanies come in the form of less-than-stellar customer feedback, a product cost that's too high, technological challenges, a changing competitive landscape, etc, late in the development process. This is very difficult to recover from since iteration is relatively time-consuming and expensive compared to software, and in many cases teams succumb to the pressure to ship something even if it's not great, or else they run out of time or money.

What would be a better approach?

My company has been pushing agile approaches to hardware development, and we see a lot of success in this model. There are quite a few challenges to applying agile development techniques to hardware on the surface, but if you can get past those, the benefits are very large. The core goal of agile processes is to get experiential prototypes in front of people quickly, as well as validating everything else that's most important about a product as early as possible. Making the right product faster requires tighter integration of hardware development with other product and business teams, and the early validation of customer and business value. Within product development teams themselves, there are gains to be made with tighter integration of design, engineering and manufacturing teams, as well as all the new technical domains coming together in an IoT landscape. With the advances in technologies like 3D printing we can get new prototypes out every day, sometimes multiple iterations in only one day. In the future, we will likely be able to apply virtual or artificial reality technology to enable experiential prototypes for people without actually making a physical product. There is a lot of room for innovation here. Agile hardware development has superior results in the new product world. However, there is still a lot of resistance in most traditional companies, with the biggest obstacle being the mindset to integrate teams and develop hardware products in this way. Furthermore, most hardware engineering is still done by in-house teams, and those teams typically aren't positioned or incentivized to share experiences around new approaches more broadly.

As we speak, what are the three most urgent topics to be tackled in the smart product world?

Product security comes to my mind; it is a really big deal in any IoT landscape. It is very easy to overlook all the different ways your product needs to be secure. We as well as our clients have to build more skills in product security; that is why we work with outside security auditors to review potential weaknesses. We also increasingly need tightly integrated technical teams who can decide where the intelligence and processing power should reside – is this in the product, in the smartphone, on the edge or in the cloud? This is a really hard question as new technologies and platforms rapidly emerge if technical teams aren't working extremely closely together. As an example of this, to maximize battery life and performance of a battery-powered product, the product architecture, hardware choices and software development all interplay with one another.

Amazon's Alexa has made voice assistants very trendy; it is one of the hottest topics right now. Do you see a wider circle of use cases for voice as a means for user interaction with products outside home speakers?

It is true that voice provides a unique and efficient interface with a product and many hardware companies are therefore looking at voice today. The advances in voice technology are good enough for usability in many consumer applications. But in my view, this is too technology-driven and not user-driven enough at the current time. In other words, the technology is still searching for the right applications for consumer markets. In some cases, the smartphone and voice control are directly competing as the interactive interface to a product. A smartphone is also a very easy interface, so I am unsure on how each one will win out with which product experiences, though I'm eager to see how this evolves!

Smart and connected products can form formidable product platforms. How do you see the relationship between the two concepts?

Platforms are without any doubt attractive from an economic perspective for businesses. But it is very difficult to make them work without creating them with a compelling consumer value proposition, so creating a winning platform amounts to a very difficult challenge. My sense is that you need to focus on user value, and develop that as fast as possible. Use whatever you can to push the adoption among prospective platform users to then enable you to have the support for an ecosystem. The Adobe Ink iPad stylus we developed with Adobe comes to mind. Adobe wanted a mobile-friendly workflow for designers, and a platform for third-party mobile input peripherals like a stylus to enable it. Their view was that people cared about compelling drawing experiences, and Adobe needed to lead by example with the new experiences they could enable, rather than only the platform. We created physical products (a stylus and digital 'ruler') and a drawing app to show what was possible with the platform. Too many companies have a 'build it and they will come' mentality with platforms.

As products get connected, do you see a trend towards changing the business model to 'as a service', perhaps with pricing based on consumption?

Let me say this. I host small dinners with many of the hardware innovation leaders in Silicon Valley, and at our last event the consensus amongst all the attendees was that hardware as a stand-alone business model is dead. There may be exceptions to this rule, but most forward-looking innovators are not thinking as hardware players alone anymore. Instead they focus more on services. But there are of course some examples of this backfiring. There is a network-connected video camera maker called Canary which started out with a traditional stand-alone product with no monthly costs. Eventually they decided to put certain features of their devices behind a paywall into a subscription model. Consumers were upset with these changes and it impacted the brand negatively. So meticulous studying of the market in advance helps a lot here.

You regularly speak about agile hardware development practices at Stanford's School of Engineering and you lead a series of discussion groups for the world's leading hardware development practitioners. How do you see the role of product engineering changing over time?

Today there is still a big gap between engineering and product development. Product development is not taught in engineering schools and most engineers are developing products in ways that seem rational and logical, but that don't benefit from the many lessons learned developing all the products that exist in the world. Furthermore, since most product development teams are still in-house, they generally lack the incentives and visibility for thought leadership in product development outside of their companies. To address this there are drivers of change needed in engineering and product development education. My dream is to share what we've learned applying agile techniques to hardware to enable every team developing a product to spend more of their time developing things that matter to people and less time on things that don't. On a cross-functional level, a tighter integration of engineering with product and business strategy teams is needed to make the right product, faster, rather than the wrong one. On a technical level, tighter integration across disciplines is needed in an IoT landscape – say hardware, software, applications, smartphones, IoT platforms, security, and network connectivity – to make a successful connected hardware product today.

Products are becoming more and more connected, dotted with sensors and loaded with on-board intelligence. This trend is in full swing. But do you see it coming to an end at some point?

The answer is a wholehearted 'No'. What we're witnessing is just the start. Consider network connectivity and 'smarts' being a tenth the cost it is today, with 10 times the performance, and how that could help enable experiences and information that improve our lives. It will be, and probably sooner than we think. Things like cars and domestic appliances are becoming more connected, but with relatively rudimentary functionality. But the idea of IoT is in its infancy. If we think of what can still be done to make objects more intelligent and data connected, this trend still has a long way to go. Fast-moving technology is the driver here. Just consider what quantum computing and blockchain will mean for the intelligence levels of today's smartened-up products once these technologies become affordable mainstream components in making a product. So, no, there is no limitation to the trend towards connected products, and connectivity will become increasingly woven into the fabric of our existence.

Note: Accenture acquired Mindtribe in August 2018

Amazon

Marco Argenti is Vice President Technology at Amazon's Web Services AWS unit. His team engages with developers and product-making companies to get Amazon technologies embedded in their solutions; he also focuses on co-developing new products with clients.

'Tapping into third-party platforms can be advantageous versus building all of these skills internally.'

More and more products are becoming connected. There is more sensing and more intelligence put into devices. What are your perspectives on this trend going forward?

We live indeed in an interesting moment in time. Multiple disruptions are happening in parallel that will change the landscape of winners in many industries. Just think of the almost unlimited computing power available for all companies regardless of size. But there are also higher-order compute services components available to practically all developers and product companies. There are more powerful networks available than ever, with high bandwidth and very low latency. And finally there is this massive improvement in machine-learning algorithms. The combination of these disruptions removes historical entry barriers. I call these effects the mass-democratization of industries.

Overall, there is major transformation needed for entering the world of connected products. What are the key challenges that incumbent product-making companies are facing?

It needs a major CEO-level decision on whether the landscape is going to fundamentally shift in their industry, given these trends. They need to answer the question, 'Do I need to become a technology company?' For some or many the answer should be 'no', as the probability of success is low. Instead they can partner with players like Amazon who make it easier to transform their products. Tapping into third-party platforms can be advantageous versus building all of these skills internally. Executives also need to decide where intelligence should reside – in the product, at the edge or in the cloud? This is a major decision. There are many examples of companies moving product intelligence to a cloud partner. For example, Valmet, who produce papermaking machines, choose to leverage Amazon's web services platform rather than building all their own skills. Another example is Pentair, a water filtration company. They use IoT analytics to optimize operations and return. By putting their intelligence into a third-party cloud, they avoid the need to hire data scientists and developers to integrate into quality processes at scale.

Will devices themselves be more intelligent in the future or does more of the processing power and intelligence move to either the edge or the cloud?

This is a very interesting question. On the one hand, today everybody can have very powerful devices for limited cost and this trend will continue. On the other hand, cloud and edge computing power is removing the need to have local computing power. Most devices will become more intelligent. But we will also see both fairly basic, low-cost devices whose intelligence is located in the cloud and also 'hub and spoke' models with smart, inexpensive devices linked to one or more powerful devices nearby.

What examples are illustrating this?

Take Amazon's Greengrass software that runs on a central gateway device. It is capable of running functions across other devices without the need for them to have their own local computing power. So, for example, in the detection of distracted drivers, a simple camera could send a video feed to a cellular tower where the powerful processing sits and works. That makes things efficient.

Do you think that voice will play a huge role as user interface in the future, now that home speakers have made this very trendy?

In the world where everything is connected, there is a need to dramatically change the user interface away from keyboard and touchscreens. Voice is the most natural form of interface and so much easier than an app on your smartphone. I guess we are soon getting to a point where customers will simply expect this capability. But there will be, in the interest of good user experiences, more than one winner across the spectrum of interfaces. Multi-user experiences and multi-channel functionalities will be mixed, so will the interfaces. Take the example where somebody books a vacation. This person might start by traditional browsing on a computer; they might then give a voice command to Alexa to find flight information, then augmented reality would visualize the location and a user app on a smartphone would track the flight progress. This requires different devices and different user interfaces to operate together. From a technological point of view two characteristics are to be taken into consideration. User interfaces need a real-time connective layer in the cloud, but actions can be initiated and completed throughout a variety of devices. These actions are becoming socially connected but also technologically linked across networks and devices.

Google

Rajen Sheth is Senior Director of Product Management, running the Google Cloud Artificial Intelligence and Machine Learning product lines at Google. A key part of his role is to help product companies embed Google AI technologies in their products.

'Every product will utilize AI within the next decade.'

Can you share your perspectives on which types of product should leverage AI and which specific AI technologies will be most relevant to product companies?

I believe every product will use AI within the next decade. The rate and pace of technology advances in this field are impressive, and I anticipate several areas where product companies will leverage AI. It will certainly play a big role in computer vision and image recognition. It should also become big in language, both text and voice as well as in translation. AI will dramatically increase its capacity to understand meaning and identify sentiment, which will be crucial for products dealing with conversational speech, where understanding intent and giving adequate response is needed. Then I think that structured data, both individual and collective, will be a big area for AI.

A lot of focus in the product-engineering world has recently been placed on voice as a user interface (UI) for devices. How common will voice as UI become in the future?

I am convinced that we will see a lot more devices with voice as user interface. That said, not all products should use AI-controlled voice interfaces by default. It is critical to start by thinking about the experience and how the user will be using the product. Once this is clear, the second step is to figure out if voice improves that experience and if it is really needed. Some tasks are definitely more naturally and more easily done via voice. For example, a Google Home device uses more task-oriented commands like 'schedule this appointment' or 'play this music on this device' than a traditional browser-based search.

A key question for product companies is to determine where the AI intelligence and processing power will sit in the future – at device levels, at the edge, or in the cloud. What do you reckon?

It is clear to me that there will be multiple layers where intelligence can reside. It can sit at sensor level or at product level. It can sit on an edge device such as a server in a retail store or on a mobile base station and of course it can reside in the cloud. The answer to where AI intelligence will sit will vary by product and also over time. I believe we will gradually see more AI on the edge, but at the moment the processing power for many of these solutions is too weak to handle execution on the edge in an accurate way. But chip technology is advancing quickly. For example, Google launched the Edge TPU – standing for Tensor Processing Unit – which is a very tiny chip, but it can do incredibly fast processing of AI applications.

The shift from traditional product to AI-enabled product means a major transformation for product companies. What are the key skills that need to be built and what are the challenges that companies are facing?

Product companies need to build AI skills, there is no question. But this is unfortunately an area of massive skill shortage globally. We estimate that there are only 10,000 specialists today that can build a world-class machine learning algorithm, while there are 23 million developers globally. Google is working to make it easier for those developers to leverage AI, so it will become less of a hassle to get these skills over time. But there is a real gap today between demand and supply of these experts. So-called 'data thinking' is also critically needed. Thinking about what kind of data exists today, what data you would like to capture, and how to structure this data is an expertise in its own right.

London Business School

Michael G. Jacobides, Professor for Entrepreneurship & Innovation at the London Business School, isn't a big fan of one-size-fits-all business strategies. Still, getting close to the final customer and making oneself as hard to replace as possible are blanket recommendations he has for businesses who venture into the complex world of smart connected products.

'The rise of the smart connected product is part and parcel with the development of seamless ecosystems that make life customized to user needs.'

Is developing smart connected products always the silver bullet for incumbent or even upstart businesses?

In a textbook world it might be the case, but reality does not support that throughout. I think that the intelligent product can principally be commoditized in a similar way to an old-style product. That is why I would caution against making your product intelligent at any cost. It might not turn out to be defensible enough. Keep in mind that your return on capital employed must be positive, at least in the medium to longer term. The customer value, your margin and perhaps those of partners have to grow substantially in order to make good on the investments you have committed for these new business lines. So, there can be no guarantee that this works out in every case.

What is the alternative for so many manufacturers whose products are attacked by disruptive competition?

I am not saying that one should stay clear of smart products. By making your product intelligent, you can of course increase the bar of competition by a certain margin and you can use the product's intelligence to produce valuable data. That might set you apart for a while in a market that is under threat of being disrupted. However, launching smart products must not necessarily offer the biggest financial benefit to you. The owner of a platform the product might be connected to, or a system integrator filtering your product into a wider solution that he might want to sell on, all also have aspirations to make their numbers add up. So, it is worth bearing in mind that you will not end up in any case with the largest value chunk from your reinvented product. Making products smart and connected makes business models complex and often financially more fragile.

Against these findings, what product set-ups work best in reality in the smart and connected world?

It differs and it is often surprising what flies and what does not. A medical inhaler that you made smart enough to measure your blood response and give you the right amount of active ingredient, can be three or four times more effective than a conventional version of the product. That is creating a lot of value for a user. However, it depends how this technology is provided to the final customer. It might be more effective to join forces with Google or Apple who have patented observational technology in the medical field and might offer a ready-to-use platform for your technology. Or take the responsive products in the automotive sector. As we speak, it is far from clear there if the smart product itself will be revolutionary or whether it will be the ability of the business to change from the provision of automobiles as products to the management of mobility needs of individuals. Oddly enough, some things that may come across as less sexy at first glance may eventually have much more disruptive and value-creating ability than initially thought. And vice versa; some great opportunities may never be realized because the necessary complementary set-up is not there for the whole system to work.

You are talking about ecosystems here, I guess; complex partner networks knit around smart products where businesses can find mutual benefit?

I think that the rise of the smart connected product is part and parcel with the development of seamless ecosystems that make life less stressful for users and more customized to the needs of the individual. I see an ecosystem as a group of specialized firms that link up for the development of something that adds value collectively. Because being smart as such usually doesn't yet constitute a value in products. But being smart in concert with other components or services that turn smartness into something that is easy to use and easy to digest is something that drives value. Most smart products would probably not yield financial successes without being woven into an ecosystem.

How do you assess the relationship between ecosystems and platforms in the context of smart connected products?

As it looks, platforms, such as the operating system of smartphones or a mobile phone bandwidth standard, are developed more often than ecosystems. Sometimes platforms do not spawn ecosystems, as they have no concomitance between their members. To me the defining difference between a platform and an ecosystem is that in an ecosystem you care about the development of things that exist in the rest of the system, whereas in platforms you might not. In that regard, ecosystems create some more specific and non-generic relationships, which essentially means that you have a slightly more closed system that sets itself against other closed systems.

Does that mean platforms are less strategically relevant for smart connected products?

No, I still think that platforms are important. The fact that they are based on much looser links can end up with them becoming generic and possibly regulation will push them to become even standardized. If these become standard for every user – an example is the 5G wireless standard – I do not think that those platforms are strategically and economically meaningful anymore. But be it as it may. I think smart products require interconnections to work. The interconnections they get are either generic and provided by platforms or they are going to be more specific, meaning that they will connect in a more seamless way with other providers that integrate them and affect either the corporate value or the customers' quality of life. The latter ones, I believe, will make the biggest impact commercially, I guess. There is a clear connection in my mind between the smart product and the platform and ecosystem dynamics as well as the strategic issues that firms have to face as they compete with other platforms and with ecosystems.

Ecosystems and platforms are complex structures to navigate for makers of smart connected products. At which end can the most value be extracted?

It differs. My own view is that value can determine who is able to create a bottleneck – by bringing an exclusive component to the table for the whole system of partners. Ecosystems are complex and often not very transparent. So you are best positioned in them where you have the final customer in sight and where the customer sees you. Then you should work towards a status where you are very hard to replace. This will be very important, because not only will you have competitors in your product segment, on things like whose product has more functionality or is more adaptable, but you also compete against the platform owners and system integrators and other third parties providing services on your product. So, competition becomes more interesting and more layered and you as a business have to become more indispensable to the final customer and less replaceable in your ecosystem. In all this, it is not necessarily a better position to be the hardware creator versus being the mere system integrator.

So, what first steps can be recommended to businesses wanting to enter their markets with smart connected products?

Having a smart product gets you in the game. But you are entering a new sphere and you should have a plan for how to deal with people that complement you in the ecosystem where the value for the final customer is jointly created. I would start with a set of questions and a decision tree to make your own position clear. Begin with asking whether you know what your smart product or service can do for the user. Are you able to defend your position within the market, yes or no? If yes, do you think that you have a strategy, yes or no? If yes, do you have a set of resource commitments that show how the strategy is actually going to take place, yes or no? If yes, do you have the organizational structure that will make it all happen, yes or no? If yes, do you have the bottom-up communication that will make you responsive enough, yes or no? Decision trees are practical because by following them, organizations can know where they stand more quickly and with greater precision.

12

Reinvented products in action

To facilitate the rotation to the new product world, we have shown that business leaders need a set of core capabilities and a clear roadmap. And we have backed up this with some encouraging testimonials from other business practitioners and thought leaders who have already begun work in this area. That is what we have so far provided in Part Three of this book.

Finally, to give the topic and our related views even more weight, we now present four long-form case studies – on Haier, Faurecia, Signify (formerly Philips lighting) and Symmons. These companies are active in different industries, are different in size and scale and are based in different regions (China, France, the Netherlands and the United States respectively). But each of them offers valuable and exclusive takeaways on how businesses representing a broad range of product makers who have embarked on the journey to reinvented products. They are all targeting different new value spaces, pivoting parts of their business to the New, and working on transforming and thus reinventing their products. Studying these examples will give you, we think, that crucial extra dose of confidence that smart connected products are where the future value is, and will be a source of inspiration to rethink your strategy accordingly.

Faurecia case study: Get behind the wheel and... relax, work and socialize!

Having overcome debt and weak auto markets in the aftermath of the financial crisis, Faurecia has come out fighting. The company's new CEO, Patrick Koller, is steering the global automotive supplier to become a systems integrator of digital cockpit and passenger technologies for the autonomous transportation markets of the future.

Playing the guitar, doing yoga or cardio exercises, shaving, knitting, and more – Faurecia strategists received a surprisingly broad list of replies when they asked car users what they would do if freed from the task of piloting vehicles.[1] Watching videos or adjusting makeup were the more prosaic activities, things many drivers already do in today's non-autonomous vehicles.

In any case, top management at the company, the sixth-largest supplier of car seats, vehicle interiors and clean mobility technology, felt encouraged by these surveys. It had decided in 2016 to make Faurecia a leading provider of 'digitally enabled on-board life' to cater for the semi- and fully autonomous electric cars forecast to hit the roads in only three to five years.

'We pitched our "cockpit of the future" to the coming car markets because we wondered what people would do in a vehicle if no longer obliged to dedicate their attention to the road. Apart from the driver, the car is already "autonomous" and nothing really compelling is so far proposed to the occupants who are not driving', CEO Patrick Koller explained in an interview at Faurecia's global headquarters in Nanterre on the fringes of the French capital.

Internal digitization precedes product reinvention

Faurecia is a stock market-listed business with a 45 per cent stake held by the French car maker Peugeot PSA.[2] Having joined its seating division in 2006, Koller took leadership of the company in May 2016. His predecessor, Yann Delabrière, had successfully turned things around after the company found itself plagued by debt and dramatically cooled automotive markets in the aftermath of the 2007 credit crisis. Up to 2015, the deployment of digital technology happened within Faurecia with a view to boosting internal efficiency levels and restoring profitability. During those years, much less focus went into the creation of digital components, features and functionalities that would enrich the company's product lines.

Once in charge, Koller, a mechanical engineer by training, who had held management roles in industries such as chemicals, lost no time in pushing ambitious strategic thinking, envisioning and acting. 'That's not something I credit only to myself,' he says. 'During the turnaround we were focused on execution, with almost no resources to think about a strategy for the future. But after that it was important for all of us in management to ask: what's next? We had sold one of our activities – exterior vehicle parts – which made us debt free and allowed us to seriously envision new things.'

Led by Koller, the management board settled for the term 'systems integrator' to frame the operational role Faurecia would fulfil in the emerging digital automotive markets. It was decided that the core of Faurecia's new product lines should be a smart cockpit intelligence platform (CIP), an information technology hub serving as the main software base for the new smart cockpit and cabin interior designs.

To back the plan up, a €100 million innovation budget was earmarked to conquer what was identified as a set of 'new value spaces'. In total, Faurecia identified potential value spaces in the area of 'sustainable mobility' as well as 'smart life on board', with the latter being divided into six sub-value spaces, of which the CIP project is one.

Roughly around the same time, Faurecia also launched an internal transformation programme to improve competitiveness in the more traditional product lines. The measures are meant to provide further financial tailwind for its pivot to new smart product ranges. Faurecia aims, for instance, to achieve a 30 per cent reduction in costs for research and development by offshoring 1,200 engineering posts to India. Faster project management tools such as e-Kanban and technologies such as AI

Figure 12.1 The smart cockpit intelligence platform as core of new product lines

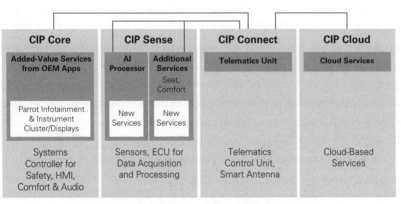

Fully Integrated Cybersecurity and AI Capabilities

SOURCE © Accenture based on Faurecia Investor presentation

have been introduced to make ideation and design processes more efficient. The company also anticipates reducing development times from 36 months to 22 months, by using blockchain technology to further increase accountability and efficiency.

In a nutshell, Koller's reinvention blueprint for Faurecia aims at transforming the company's core business and operations by pivoting their product ranges, workforce and processes to generate savings by leveraging digital technologies. These savings will be used to further grow the company's three core business pillars – seating, interior and clean mobility technology – and, in parallel, build scale in the new value spaces.

Interiors and safety turning into differentiators

Koller has put out ambitious financial targets to Faurecia's shareholders for both its core product ranges and, especially, the new smart connected product lines. By 2025 the company aims to reach €30 billion of total sales, representing annual growth of above 8 per cent. The new value spaces are expected to grow three times as fast. While smart cockpit technology is expected to reach €4.2 billion, clean mobility technology is forecast to turn over €2.6 billion by 2025, reflecting an annual growth rate of 33 per cent. In 2017, Faurecia registered an overall turnover of around €17 billion, with global car producers such as Volkswagen, Ford, Renault-Nissan, BMW, PSA and Daimler.[3]

Koller is convinced that the automotive markets will experience a seminal turning of tides. 'Once the car becomes autonomous, you will, as a carmaker, no longer be able to differentiate yourself via your brand, through the exterior design or through the specifics of your powertrain. You will differentiate yourself mainly through the user experience inside your vehicle and its interior smartness', he argues.

With no drivers left in autonomous traffic, nobody is interested in horsepower and chassis design, especially with road traffic set to turn fully electric. All there will be is passengers wanting to get from A to B as quickly, smoothly and safely as possible while enjoying maximum attractive opportunities to work, relax or socialize. Cars must therefore be connected at all times, with passengers able to use the same apps as they would at home or work. Faurecia is now the only global supplier left controlling all vehicle interior elements, so it was logical to go for smart technologies and integrate them with existing design features to deliver tailored solutions for targeted user groups. 'It was common sense for us to say: we should be a key player here', Koller recalls.

In late 2016, the CEO initiated the creation of a new cross-business team called 'Cockpit of the Future' (CoF). It attracted some initial scepticism and even slight resistance within the organization, as it was funded through existing business groups' innovation teams. In some quarters, the idea of digitizing seats and interiors was seen as nursing expensive gadgets with no real economic prospects. Still, these headwinds were overcome by arguing the opportunities of drastically changing markets, so that the plan got enough senior people on board for launch.

Systems integration needs a range of partners

At first Koller's CoF initiative merely entailed laboratory work. A team of 15 started ideating on new cockpit configurations and usages in a new venture located at Faurecia's research site in Meru, France. The team's first task was to help carmaker clients to envision and test drive new propositions for the autonomous car cockpit of the future.

As the operational leader responsible for the whole CoF initiative, David Degrange, an experienced engineer and business development manager, was appointed, reporting to a board of seating and interior division directors. After his 'lab' churned out a number of encouraging first ideas, the CoF team took on strategic responsibility to commercialize its new digital solutions.

Yet Faurecia's management saw clearly that, to become a system integrator in fast-moving digital car markets, they needed to fill expertise gaps. To this end, the company vigorously immersed itself in an ecosystem of specialized partner businesses. A few were even competitors in certain product fields and all were innovators in their own fields, sharing Faurecia's vision of the cockpit of the future.

'As a system integrator we understand the full value chain of the cockpit of the future. So we can identify points where it doesn't make sense for us to invest on our own because the entry costs are too high and because you would anyway have world leaders with stronger market positions on hand you could partner with who are recognized experts in these domains', Koller explains.

Faurecia saw, for instance, that to properly handle car safety, a development alliance with ZF, a German-based specialist in car safety systems, made strategic sense. For similar reasons, it went into a co-development deal with the specialist Mahle, a world leader in vehicle air conditioning systems. 'The typical air conditioning system in cars today feels too mechanical. It will become an electrical system to make it even smaller but also to personalize the climate comfort', Koller says.

Acquisitions to beef up software teams

But for some key capabilities, especially in embedded software and AI-powered assistant technologies, Faurecia felt it needed new expertise under its own roof to build the cockpit platform. Koller therefore decided to make bolt-on acquisitions. In 2017 Faurecia bought French-based car infotainment specialist Parrot for €100 million, and went into an equity joint venture with Chinese company Coagent for €193 million.

Both firms offer software know-how Faurecia did not have. Parrot brings strong automotive software capabilities to the table, with 300 engineers located in Paris, while Coagent provides 400 software specialists as well as manufacturing capabilities in China, commanding an 8 per cent market share as a software provider to Chinese carmakers. Under the headline 'Faurecia Tech', the company created a distinct architecture of knowledge blocks tying in not only mature corporate partners but also academics, start-ups and other technology platforms.

All Faurecia's partners share a strong innovation culture and their expertise will play a role in future automotive markets. Security and

Figure 12.2 Faurecia's technology organization to accelerate innovation and transformation

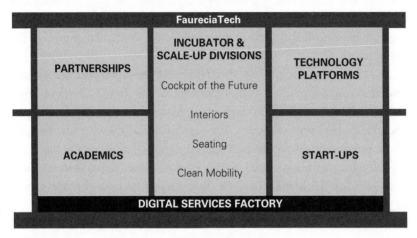

safety will be key in the adoption of autonomous vehicle technology, thermal management will be crucial for extending the range of electric vehicles, and AI and cloud computing will be critical drivers of smart on-board features such as voice assistants.

Integrating technologies

To get support in tying all this expertise together and building the complex informational technology architecture behind smart car interiors, Koller added Accenture as a partner to its ecosystem. 'Their expertise is digital transformation and IoT landscapes. Again, we are not in this field, but we want to benefit from the innovations. We do not work with different industries like Accenture does, so they have a much better understanding of what the market offers and are in fact accelerating our capacity to integrate the best solutions', he explains.

With Accenture's support, Faurecia conducted a string of ideation sessions to convert use cases into actual product solution lines for CIP.

David Degrange and his leadership team prioritized the technologies to be developed for the intelligence platform. They proposed a short list of features and Koller had the final say. Ultimately, the overall vision encompasses not just the core but also sensing, acting, and cloud computing, in a unique blueprint that is currently without parallel elsewhere on the market.

To identify these unique and differentiating product features, Faurecia undertook a few market studies starting in late 2016. The team also conducted user interviews to create 'hassle maps' and to build use cases and group them into segments. For each of the segments, they thought about offering distinctive technology solutions. Continuing on, the team is conducting further regional market studies in the United States, China and Germany. These will focus on the changing needs and usage habits of potential car drivers and passengers.

The cockpit of the future is highly adaptable

The development of the digital cockpit as part of Faurecia's 'new value space' strategy is strongly guided by user needs and expectations. The car can completely reconfigure for each individual driver and for external driving conditions. The cockpit seamlessly integrates all smart components: electronics, active decoration, smart surfaces and actuation, intuitive human-to-machine interfaces, multiple transformable touch displays, state-of-the-art software applications, adaptive seating with sensors for biometric data collection, and infotainment elements. Voice assistants act as principal interfaces to control functions. The cockpit also combines enhanced passenger safety, predictive wellbeing, and transportation convenience with a maximum of external connectivity.

For its voice control, the cockpit is fitted with a choice of voice assistants (including the popular Amazon Alexa, Google Assistant, and Baidu DuerOs) differentiating between user voices so that each passenger can direct their preferred assistant to perform separate tasks. Among other functions, the assistant can adjust seat positions, start a seat massage, change climate settings, and upload video and music playlists. Crucially, users can tap into the car's settings on their preferred voice assistant at home or in the office. They can, for example, consult and update to-do and shopping lists while in the vehicle or get their car ready to drive by setting the climate control levels from home. The assistant can also be used to make maintenance appointments with garages and coordinate them with those business's MS Office calendars.

The traditional driver dashboard has been reshaped to include one large digital display. Instrument panels consist of adaptive surfaces that are able to change display, functionality and position. The cockpit also features facial recognition, allowing the vehicle to suggest music or specific scenic routes depending on driver identity or mood. Each passenger can enjoy privacy, particularly when it comes to sound. Thanks to designated 'sound bubbles', the driver can follow GPS instructions while passengers listen to their choice of audio or have a phone conversation without anyone disturbing anyone else.

Patrick Nebout, Vice President of innovation in the seating business, says Faurecia has also developed an 'active wellness' seat, which employs biometric sensors and predictive analytics to measure and respond to occupant stress, drowsiness and other symptoms. Via a smart trim, the seat collects an expanded range of biological and behavioural data, such as heart rate, respiration rate, body movement – eg fidgeting – and humidity. For a smarter and safer seat, Faurecia and ZF developed a frame concept that allows cabin occupants to drive, relax and work safely and seamlessly. The seatbelt, belt retractor and airbags are all integrated into the seat, with these safety functions designed to operate optimally in different seat positions.

Next to its many adaptive features, the cockpit also offers predictive functionalities. It can, for instance, anticipate a driver's preferences and adjust positions of seat, mirrors and steering wheel. More important, the cockpit is intelligent enough to anticipate individual drivers and passenger safety levels and adjust controls, displays and autonomous handovers to match. This is of particular value for car-sharing schemes, where drivers have their profile and driving history saved so that they get personally preconfigured cars regardless of where they hire a vehicle.

Going via flexibility and personalization from B2B to B2C

Faurecia has in view a potential market of €35 billion for technologies associated with its intelligent cockpit by 2025. Assuming that the majority of this potential is realized between 2017 and 2022, and further assuming that Faurecia will gain a 15 per cent share of this market, this technology alone, according to analysts, could add 3 per cent per annum to Faurecia's stock value over the next five years.

The company's CEO sees its incumbent business model as changing over time as the demand for flexibility and individualization grows in the car markets of the future:

> 'You will have to provide equipment solutions that can continue to individualize and adjust to new needs. You as a consumer buy a car interior with some use cases and two years later you have a baby, so you need a configuration for a third use case. It's never exactly the same car. This degree of flexibility is not built in yet today but it will happen. And this will also change our business model, which is today a B2B model. At the moment we are moving with the cockpit to a B2B2C model and eventually we will have a significant business in B2C markets too.'

Signify case study: LED there be smart light!

Seven years ago, a band of visionary middle managers at Royal Philips pulled off an innovation stunt that rang in the smart lighting era for the lighting division, which was rebranded as Signify in 2018. Determination to create a pioneering product, expertise harvested from across the company, and mission-critical marketing attended the cradle of the Hue smart lighting system, which turned out to be the precursor of a whole new operating model.

Royal Philips is one of Europe's oldest and most venerable technology leaders. In 2016, it spun off its lighting division into a separate company to open the business up to new investors. Two years after the split, the new entity changed the name to Signify. It is today by far the world's largest provider of lighting products, systems and services, at twice the size of its nearest competitor.

Light-emitting diodes, more commonly known as LEDs, have existed for decades but only reached acceptance in lighting mass markets around 10 years ago. At Signify, compared to its rivals, the boat wasn't that seriously rocked by this disruptive novel technology because the company's engineers had actually helped this new market to bloom. Signify teams, for example, had invented the dual optics technique, enabling the use of LEDs for directed road illumination. Amongst other LED innovations, the company's engineers were also the first to introduce entertainment lighting that used multiple red, green, and blue LEDs for precise, colour-coordinated effects in theatre and architectural lighting.

Such vanguard innovations positioned the company from very early on as the market-leading provider of smart connected LED lighting systems, but more time was needed before it really arrived at its first mass-market smart lighting LED product. Industry-wide, foundation technologies had to mature first, such as the Zigbee radio protocol, which Philips had a crucial role in developing. What is more, smartphones and their application libraries had still to arrive on the scene.

With its LED innovations the company was the driving force behind the industry's largest disruption, going from conventional to LED lighting. With its LED-based lighting sales increasing from 14 per cent of the market in 2012 to 70 per cent in 2018, it is the world's largest provider of LED lighting.

Signify's smart lighting trailblazer

'If I go back only five years, conventional lighting constituted about 80 per cent of our product ranges. Now we are at around 70 per cent LED', says Bill Bien, Signify's leader for marketing and strategy alliances. In not even half a decade, he says, the newly created 'Systems and Services' businesses, which include Signify's smart lighting for consumer and professional customer segments, grew from zero to more than €900 million, or more than 10 per cent of Signify's revenues in 2017.[4]

But at first the new smart LED era needed a kick-start product, an emblematic pathbreaker with real market success to accelerate the organization's move towards smart lighting. For consumer lighting this came in the shape of Philips Hue, a system of colour-tuneable LED bulbs that can be controlled through a smartphone app, motion sensors, and connected switches. The system turns LED lighting at home into an innovative, distinguished and personalized home essential. Hue smart lighting is dimmable, timeable, and controllable through iOS or Android devices to set light intensity and colour. Individual scenes such as sunsets can be set for different rooms. The bulbs were conceptualized as open-platform digital products, and over 700 third-party apps have so far been launched for the user community to tap into. Among them are such popular apps as Hue Disco for parties and Hue Manic for coordination of light and music. Hue alone created close to €300 million in sales last year – a third of Signify's smart lighting sales.

'The idea started around 2010 when I was doing work with mobile apps in another part of Philips', explains Jeroen de Waal, then head of global marketing for Philips' lamp business. 'We thought, it is quite convenient to control lighting at home via your smartphone. This idea had been discussed in the corridors previously and I had some conversations with people from the LED department', the manager recalls.

What, with hindsight, turned out to be a true light-bulb moment started with an idea that a small group of people working as middle managers floated within the larger company. De Waal says, 'Digital and mobile apps were not big on the radar and some people asked, "Why would you try that?"'

But armed with plenty of innovator conviction, the start-up team kept repeating what they thought would be successful this time round. They argued, 'It is LED, it can be controlled with a smartphone, it will therefore be much easier for consumers and easier to market', de Waal recalls, adding, 'It is worth remembering that the smartphone market itself was likewise in its infancy at this point. Apple's iPhone, the device that led the new market to global explosion, only launched three years before Hue was ideated.'

The aim was not to come up with something fancy that ended up in a niche market but to expand lighting on a broad basis into an everyday Internet-connected smart product, George Yianni, who served as head of technology for Hue, explains: 'A traditional light bulb would have a replacement cycle of one year. Now, with LED, all of a sudden, lighting products could last much longer. So, we asked ourselves, "What kind of innovation do we need to use LED's disruptive effect to create additional value for our customers and for Signify ourselves?"'

Marketers and engineers as equal drivers of innovation

The practical first steps involved a one-day launch meeting where around 50 people from a wide variety of business divisions were invited. 'We brought all this expertise together to channel our thoughts. What would we actually know about all this?' It turned out to be quite a lot. Philips engineers were already working on technology components that could be helpful for developing a smart home connected lighting system. There was some knowledge of wireless protocols, software engineering and APIs. There were even tentative initiatives to build smart LED controls. So, the engineering ground seemed well prepared, but one element was sorely missing: a very clearly shaped consumer proposition for a personalized smart lighting system. 'What do we actually want to go to market with? How do we want to go to market? These were the burning questions. That is why we needed a very good marketer on board', says de Waal.

The team found this figure in its group of experienced Philips marketers, with extensive knowledge of consumer markets. Together with Yianni, the leading engineering and marketing minds of the project gave Hue the shape markets longed for. The marketers looked at Hue strictly through consumer eyes, determining a final list of features for the new product: it had to be smartphone-controlled, it had to work with colour and third-party apps, it needed an online distribution channel and its own social media world where users could find inspiration and build a user community.

'A colour technology had to be used for Hue that allowed us to get the emotional mood into the product and to underscore that this is lighting of a very new and revolutionary quality', Yianni explains.

Looking back, the team agrees that it was a joint effort between engineering and marketing specialists. The marketers had for instance taken the decision on the final technology stack supporting Hue. They took the responsibility for how one seamless consumer experience could be built for the product. It was crucial for them to really get a story across, conveying the emotional implications of using the product. Only that would really make it a breakthrough from the consumer perspective. It was similarly critical, as the Hue inventors recall, to pick a bold and punchy marketing claim, which they found in 'Lighting has changed', as well as to get Apple's retail outlets on board for distribution right from the start. 'The marketing part was supercritical in driving the initial success, while the technology architecture was supercritical for sustaining it', de Waal says.

Home lighting has traditionally been seen as a low-engagement category. Consumers bought a new light bulb when the one in use broke, or when they redecorated a room and needed a new fixture. The Hue team therefore decided that they would market their lighting system under new and strongly emotional headlines. 'It could be the health effects of lighting, it could be about how to beautify a home environment. It could even be about the security benefits of lighting', says Yianni. Before Hue there were no solutions for normal consumers that were easy enough to install at home or even affordable for them. Hue's new, less costly technology stacks put the team in a position to replicate for the ordinary buyer what was formerly only available to affluent clients. 'Thus, by loading Hue with emotions of self-expression and home improvement we could transform lighting into a high-engagement mass market', Yianni says.

A tiny dedicated team

The starting team around de Waal and Yianni started with only four people and grew to 30 by market launch. The project was placed in a designated area for new ideas within the company. It was about bringing together capabilities that existed across the organization, remixing them to make them agile and innovative enough to pull this off. The team had approval to hire and also secured the backing of the head of research and development to commit resources.

Back in 2012, de Waal admits, their disruptive proposition must, from any perspective, have appeared risky, having no guarantee of market success at all. In making R&D investment decisions, some leaders would always have found incremental innovations they preferred over Hue. However, Philips always has been willing to take risks with new technologies. In the end, Hue was born within a highly innovative and technologically apt company, as Signify's ample pedigree in LED innovations demonstrates.

Marrying solid expertise with entrepreneurial propositions

When the Hue team started developing, non-smart LED lighting was already big business for Signify; it was essentially about taking a traditional light bulb and turning it into an LED, making it more efficient, yielding better light quality with greater value for end customers. 'We had this capability and this very high level and quality expertise on LED lamps. Philips engineers knew everything about manufacturing and research and development of these products. And on the other hand you have us, this venture, where you have people with lots of wild thoughts and propositions but no experience with scaling, manufacturing or quality management on products. We had to marry these two things successfully', de Waal says.

Hue's model sets the tone for smart lighting

And they did. The early success of Hue was part of Signify's intent to systematically push forward with smart lighting in many other aspects of its product portfolio, especially for business clients. At the same time as the Hue launch, Signify was already developing and offering professional smart lighting systems such as Color Kinetics for dynamic architectural lighting, and CityTouch for smart street lighting.

Over time first Philips and now Signify have expanded these systems and today offer multiple smart lighting architectures for various systems that comprise digital LED and IoT technology. The company has, for example, created an indoor positioning system based on lighting for retailers that allows hyper-accurate navigation to retail products. Some of its other lighting systems provide services where occupancy rates in office buildings are established and assets located on manufacturing shop floors. Signify has also launched the Interact portal, which provides data-enabled services back to its customers.

Signify is keeping its foot down on the innovation accelerator. It has put a strategy in place to bring an innovative smart LED product to market every six months, not just for the home, but for industry, retail, hospitality and public infrastructure clients. The company has a start-up mentality and regularly hatches new ideas for LED applications. Hue was an early venture but there are more. For example, its horticultural lighting group offers systems where tuning of LED light quality and temperature leads to better growth rates in plants. Another is called 'Interact City', which provides smart street lighting and has created solutions for over 1,000 cities.

The Hue product started out with some visionaries scribbling on a piece of paper, and Signify had the means to give scale to a pioneering idea. This can be the model for many other product companies.

Symmons case study: Turning pipe dreams into a digital business

Having an 80+-year-old business with a strong innovation legacy in conventional plumbing products, the US-based Symmons Industries has developed Symmons Water Management to create a new category of connected plumbing. It is a smart hardware solution which will solve the pain points of poor hot water experiences in hotels through a network of sensors, data aggregation and analytics – to create optimal guest experiences for travellers around the world.

The run-up to the product reinvention

Showering used to be a painful experience... quite literally.

In 1938, while working for a small plumbing manufacturer in Boston, Paul C Symmons developed the idea of constructing an efficient pressure-balancing valve to provide a safe showering experience. Symmons speculated that the problem of drastic and unsafe temperature change in the shower is caused by a fluctuation in water pressure which occurs as a result of demands elsewhere in the plumbing system, such as turning on a tap, flushing a toilet or running a dishwasher. Scalding was commonplace....

Realizing that there was a need for a product that would improve the showering experience was the motivation for Paul Symmons to start his own company in 1939 and design a valve that not only solved the problem of water temperature fluctuation, avoiding scalding, but also revolutionized the plumbing industry. The introduction of this unique Temptrol valve, coupled with favourable plumbing code changes, helped Symmons Industries to grow rapidly over the next 25 years, where the product remained virtually unchanged. The growth of the company was managed under the direction of Bill O'Keeffe, Paul's son-in-law. In 2008, Symmons rebranded itself by expanding its product line to include premium kitchen and bath products. The product expansion, as well as the new approach to selling, enabled Symmons to reassert itself in the plumbing industry.

In 2010, Tim O'Keefe became the third generation of the family to take the reins of Symmons Industries after a career in the fast-growing enterprise software business. Tim brought an interesting DNA to the business as a visionary and strategist. He set out on a journey to digitalize many aspects of the company's core operations. He also began expanding Symmons' focus deeper into hospitality markets, achieving an impressive market share with hotel properties across North America, building the company's first design studio and creating a broad range of functional and aesthetic options in designing bathroom fixtures.

While Tim created many improvements in the business, his entrepreneurial passion continued to 'flow'. Tim recognized his customers had needs that could be addressed through new and evolving technology. This led to a new charter for Symmons; a charter to enter the smart building space. Tim's central belief was that water was one of the few systems in a building that is 'dark', unlike electrical and HVAC systems. He believed that with data, many problems could be addressed, and new business opportunities created. But the question remained, where to start?

The road to (successful) reinvention

In 2017, Symmons engaged Altitude Inc. (now part of Accenture) to take a different path to innovation. In this case, the engagement was not starting with a pre-defined solution and building from a set of existing assumptions on what customers (and hotel guests) wanted, but rather with a discovery process to uncover deeper insights into Symmons' current and potential new customers. A project was chartered which engaged with over 100 hotel property managers, plumbing engineers and building owners over several months. Alas, truths were beginning to be revealed.

What Symmons learned in this process was that hotels did not seek to differentiate in the bathroom and bathroom fixtures specifically, where Symmons' product development was focused. Hotels indicated that they wanted to invest in the 'buzz'-worthy common areas where people spent time together and where every dollar of investment would be seen by more eyes. The bathroom, in contrast, was a private place, not often shared on social media… quite simply, property managers wanted the bathroom to just WORK! From what the team learned through site visits, a working shower was not always the standard. Based on these qualitative insights, deeper study was commissioned to understand the frequency and impact of plumbing problems in hotels. After reviewing 800,000 social media records, Symmons could now confirm what their customers were saying, with hot water as the #2 pain point for hotel guests, right behind Wi-Fi connection. As it turned out, water did not always work – it was unreliable, the source of the problem often unknowable, and from a reviews perspective, it was uncompromising.

From this insight into a bathroom that works, the Symmons Water Management lighthouse vision was born. Symmons had found their 'because we should' moment, a clear pain point that was impacting their customers' reputation and business. The discovery process sought to understand customers' unmet and unarticulated needs more deeply through ethnography. With this new insight, Symmons identified a vast opportunity space for innovation that would solve this core pain point and bring substantial growth to the company. The question remained, where to start?

While a lighthouse serves as the beacon for long-term ambition, it remained critical to back-cast steps to that vision. The first step was to go where there were the highest number of problems with the fewest addressable solutions. A concept was born: Symmons would monitor hot water systems on behalf of hotel properties. By simply connecting components of the building's plumbing system, they could monitor three crucial water variables: temperature, flow and leak detection. With this data, Symmons could meet their minimal viable product (MVP) concept to detect problems earlier, trigger corrective actions, and ultimately improve the hot water experience for guests.

From this first MVP concept, Symmons continued the customer-centric innovation process with their partner and chartered development of a proof-of-concept solution to be built with a series of agile sprints. This first generation of Symmons Water Management included off-the-shelf sensor hardware and network gateways, an IoT application in the cloud, and a data analytics platform which would put plumbing system performance data in the hands of property managers and plumbing engineers. The key was to get real-world experience and here Symmons commissioned this proof of concept to be deployed with four large hotels in the Boston area to not only provide insights into the technical solution, but to begin testing the multiple elements of the value proposition.

Commercializing the new business

'What Nest did for thermostats, we're going to do for water. We believe that products should do more for you than you do for them.'

Tim O'Keeffe, Symmons Industries CEO

Armed with significant data from the proof of concept and validation of their customer value proposition, Symmons is poised for a successful first product launch in 2019, but this remains a story in progress. Symmons is going to market early in 2019 and will use this platform as a way to continue learning and perfecting their offering. As Symmons Water Management is creating a new market for connected plumbing, a very different approach must be used to commercialize the business when compared to an innovation with an existing or mature market. This requires building awareness of the problem/opportunity among key customers, working with them to understand the value proposition and piloting in close cooperation to assess the value for the property. Once this basis is established, then a broader push within the industry can be made to set a new benchmark for hot water experiences in hotels, where hopefully hotel guests will demand a Symmons Water Management shower.

The path to creating this new business contains many pitfalls beyond just product–market fit. How to scale? How to maintain focus on both the core business and new business simultaneously? How to build an innovation culture and digital competencies in an 80+-year-old industrial product manufacturer? Symmons understands the focus and dedication that establishing both a new market and business requires. They have appointed dedicated leadership to Symmons Water Management, building an ecosystem of partners for a new digital product, new go-to-market strategy, and very different internal capabilities such as software development, hardware engineering and product support. They are carefully balancing the current and the new, bringing the core organization along with them, from inspiration and vision of Symmons Water Management to execution, without distracting them from what needs to be built today.

What Symmons has learned, and is putting into practice every day, is that innovation can only be impactful if there is a defined customer need behind it. It may not be an obvious or acknowledged need, but once that need has been excavated, the path to solution development becomes clear and exhilarating. Keeping customers tightly looped into the continual development of each successive Symmons Water Management offering has become a core requirement for Tim and his team at Symmons.

Haier Group case study: Putting a platform before the palate

Home appliances group Haier Electronics is pioneering new management methods focused on instigating digital entrepreneurship, innovation and the creation of business platforms. These are stress testing the company's maxim 'the user is right'. Haier impressively shows how traditional operating models can be pivoted to a new life by turning to completely new appliance ranges that are designed, made and marketed as platform products. By doing so, a unique digital technology stack is embedded into entirely new or heavily re-engineered legacy appliances, while this is accompanied by drastic deregulation of hierarchical non-agile management structures. Two smart connected products stemming from Haier's cooling range provide evidence that this approach, entirely new to the sector, works well. They have, in fact, earned the company the position of world leader in its market. Digital maturity in China is famously high. Serving a sophisticated domestic consumer market, AI and other smart technologies are reinventing China's products more than almost any others in the world. Eric Schaeffer, co-author of this book, enjoyed the privilege of being invited by Haier Group to its Chinese headquarters, where he met the company's restless digital reformer, CEO Zhang Ruimin.

Chinese Haier Group refers to itself as 'the world's fastest-growing home appliance company.'[5] On the back of various recent acquisitions, it has grown to become the biggest brand in the global appliances market, currently controlling a 10.5 per cent volume share.[6]

A sprawling business organization, Haier has been led for more than three decades by CEO Zhang Ruimin. He is seen as a visionary who has reorganized the company's once sclerotic hierarchies to make them more agile for the digital age. Central to Zhang's thinking is the creed that an appliance maker, in order to survive the disruptive forces his sector faces and succeed in a digitally enabled consumer world, has to operate not only as a platform business but also as a creator of product platforms – rather than just being a traditional hardware product manufacturer and manager of brands.

Having decentralized its organization from its home base in China to operate all around the globe, Haier now runs 66 trading entities, 10 research and development centres, 108 manufacturing locations and 24 innovative industrial parks, dotted across all major continents. The group employs over 70,000 people and in 2017 its global revenues stood at RMB 241.9 billion (around US \$36 billion), representing a 20 per cent leap from the previous year.[7]

Headquartered in Qingdao, Haier designs, develops, makes and distributes a wide range of domestic products such as refrigerators, washing machines, microwave ovens, televisions, air conditioners, mobile phones, and computers. Each brand and product line has its own market position and provides, to a larger or lesser extent, smart home experiences to users. The brand portfolio includes names such as Haier, Casarte, Leader, GE Appliances, AQUA, and Fisher & Paykel.

Bottling the innovation opportunity

One of Haier's most digitally advanced products is a 72-bottle wine-cooling cabinet, enabled for connection to the Internet of Things (IoT). The product is designed for use in private households, restaurants and wine merchants and, as a smart product, it acts as a platform around which an ecosystem of partners can be formed.

Also from Haier's range of cooling products, the company's intelligent 'link cook' refrigerator is another smart product designed to become a domestic product platform in the future.

Both of these devices derive from close studies of Chinese consumer trends. In the case of the wine cooler, detailed market research found that the peculiarities of the Chinese wine market allow for a hub appliance that not only stores wine bottles in adequate temperature and light environments, but also shares information and data and facilitates commercial interaction between wine makers and wine drinkers.

Sino vino: towards a mass market for wine in China

Although wine has been cultivated in the region for more than 1,000 years, China has not yet become a wine country in the western European sense, with established mass markets, an authoritative wine guide culture and consumers commanding the necessary purchasing power. China's current wine consumption still only stands at 0.4 litres per person per year, compared to France's 50 litres.[8] But Chinese wine consumption is likely to grow commercially, driven by the country's economic boom and the rise of a solvent consumer middle class. Overall demand is estimated to reach US $23 billion by 2021[9] and, on the back of its huge population, the country has already become the world's largest red wine market.[10]

Wu Yong, director of cooling for the China region at Haier comments, 'Despite all this, a really sophisticated wine culture is only just emerging in China. Consumers have not yet gathered much experience with the enormous breadth of wine types on offer. There are literally thousands of brands and labels in many different languages available and prices vary widely.' He adds that most wines consumed in China are still imported from abroad and that there is a need for transparency across the various wine categories, origins and varieties. There is, in his view, also room for background education on wine making and wine consumption: 'We want to create something useful for the wine producer and the retail merchant as well as for the wine drinker. We want to foster the creation of a community that shares the lifestyle of wine drinking.'

Haier's wine cooler cabinet boasts some cutting-edge functional features that make it stand out from the crowd of competing devices. As one of the first of its kind, it works with a compressor-free cooling technology, reducing wine-damaging vibrations to zero. The unit offers different temperature compartments for various wine types so that reds and whites can be kept in an adaptable range between 5 and 20 degrees Celsius. The cooler's glass front is also designed to filter UV radiation, protecting the wine from light damage.

Though all these features make the product a really advanced piece of appliance engineering, they do not yet render it an intelligent device, a status only conferred by the sophisticated IoT features built into the cabinet. These enable it to analyse data, connect to the cloud and use artificial intelligence to deliver true consumer interaction. The parallels with devices like Apple's smartphone are obvious. The iPhone was a highly engineered and innovative box that gave them the competitive edge even before its platform qualities came into their own.

Building a tightly knit community around wine consumption

As its main interface with users, the smart cabinet carries a 21.5-inch touchscreen, a loudspeaker with voice recognition capability and a camera. This allows wine drinkers or delivery personnel to scan and log bottles going into the cabinet, and lets restaurants and private households keep track of their stock. Each bottle identifies itself to the cooler via an RFID chip carrying information about taste profiles, originating vineyards and supply routes from producer to consumer.[11] The cloud-based database storing this information, says Yong, holds, at any given time, a minimum of 600,000 wine data points available to consumers. Thus, wine drinkers can also scan bottle labels to get them analysed for background information or they can use the screen to make the cabinet suggest wines for certain meal types. Sitting right above the glass front, the screen can also educate users via videos on wine making and advice on the best drinking experience.

Cutting out the middleman with a platform freebie

Bottles can be automatically reordered via the interface from a favoured and trusted wine producer, cutting out intermediaries such as wholesalers and importers. Vintners and wine makers, in turn, can use the display screen to show their latest offers and run targeted promotional campaigns based on consumption data shared by the device. So, a wine producer located at a chateau somewhere in the French Bordeaux region can directly tap into the habits of wine drinkers based in Shanghai. Chinese wine drinkers, likewise, can get in touch with favourite winemakers two continents away for suggestions and direct feedback.

Haier crucially positions the product as a platform device. In a major departure from traditional appliance maker business models, the company gives the cooler cabinet away free. For income from the product, Haier signs contracts with its ecosystem partners: wine makers, importers, restaurants, logistics services and, of course, winemakers. These arrangements stipulate that a certain proportion of the total turnover created by the devices – in the shape of wine bottles bought and reordered by consumers and restaurants – lands with the manufacturer of the cabinet.

In a country where red wine consumption stands at around 1.86 billion bottles a year,[12] this can be a viable business model, especially if wine is only just emerging as a mass consumer product. Still, Wu Yong acknowledges that the model is not yet profitable, although it is expected to be soon. He says it takes around one to two years to recoup the production cost of each wine cabinet.

Haier's particular platform approach to this product makes it necessary to connect to and maintain relationships with local business entities on both the supply and demand sides of the wine market. This confronts this giant organization with the challenge of often complex and extremely granular supply, distribution and relationship management – for instance with wine producers in Europe and Australia. This was one of the reasons, Yong says, why Haier decentralized its organization, thereby achieving better product management quality and greater financial success with its devices. Yong says, 'It is not easy to achieve. But we saw it as a big plus in terms of consumer experience. The platform enables wine drinkers to get their bottles at a lower price, as our platform model cuts out the middleman. And wine drinkers get their favourite brands quicker via the direct relationship network the platform creates.'

Making a fridge the social hub of the house

Like the smart wine cabinet, the Link Cook refrigerator, Haier's second flagship smart product, was conceived on the basis of meticulous consumer research. Like the wine cooler, this device works with a large touchscreen on the outside serving as its main interaction point with users. Crucially this appliance is connected to Haier's U+ Smart Home platform, through which the manufacturer orchestrates its range of kitchen appliances.

The smart refrigerator connects through this platform with Haier's micro ovens and cooker hoods, forming a seamless production line for meals. It is sensorized and camera-equipped so that it can identify, via intelligent algorithms, what its current content is and which items carry which sell-by dates. On this basis, the refrigerator, when consulted by consumers, not only presents an up-to-date shopping list of things that need replacing, it also suggests recipes using the available ingredients, taking into account freshness and expiry dates. 'The device then automatically prepares the micro oven for action by setting the right temperature and cooking time for the meal chosen', Wu Yong explains. Its recipe then appears on a separate screen on the range hood so that the chef can consult it while chopping and mixing ingredients as advised.

The touchscreen also allows users to check the weather, deal with e-mails or even watch television. Haier's refrigerator concept aims at matching and boosting various trends in meal preparation among younger generations in China. Though digitally savvy and fascinated by automated experiences in day-to-day life, this age group has often lost knowledge on recipe ingredients, cooking styles, food handling and preparation methods. 'External assistance such as culinary education, tips on food management and healthy nutrition as well as tutored meal preparation are therefore much appreciated by this target group', says Yong. What is more, the kitchen – at least in Asia – is becoming ever more the main social hub of family life. This creates the opportunity to combine meal preparation with socializing, leisure activities, information exchange and digitally supported interaction with the outside world.

A community platform in the waiting area

Haier sees its smart refrigerator, unlike the wine cooler, as being still in an early phase on the path to becoming a fully developed platform product. Where the wine cabinet is provided to consumers as free hardware, the smart refrigerator is sold like a traditional domestic appliance. However, its developers have, from the start, conceptualized the device as a potential platform by equipping it with on-board intelligence and connectivity features that, at a later stage in product development, can serve as anchor points for ecosystem partners.

'We seek to transform all of Haier's domestic appliances eventually into "smartphone solutions" that go beyond isolated functionality into a connected experience at home', says Wu Yong. That means that just as Haier's wine cooler already links up a community of wine drinkers with merchants and producers, the smart refrigerator could one day establish a 'living' real-time relationship between food producers and consumers, and cut out intermediaries. For Haier, this would then mean establishing a network of partners among food producers, consumer goods manufacturers, and logistics partners with whom the appliances maker would then enter into revenue-sharing deals.

This is an energetic push into the smart connected product world, in line with Haier's declared aim of eventually selling nothing but intelligent appliance ranges. However, it also shows that a very traditional legacy product line should not be changed beyond recognition to achieve platform building and service provision. The pivot from old to new might be technologically massive, but it can build on solid know-how in hardware engineering and this will ultimately help drive success in the digital.

Accenture has conducted research into the Haier products under the GE appliance brand – dishwashers, ovens, refrigerators, dryers and washing machines – and their competitor peer group of connected appliances. This shows that Haier's connected products achieve not only the most versatile value drivers, such as hyper-personalized experiences and third-party ecosystem experiences, but also the highest value creation overall. Even with a smaller number of connected products, GE offers a greater spectrum of digital services than its peers. This must be one of the driving forces behind its recent sprint into the position of world leader in this market.

Newly won agility delivers speed to market

Both products, the smart wine cooler unit and the smart refrigerator, have been brought to market in just over a year – an extremely short time. This impressive speed from ideation to retail has been made possible by extremely agile organizational structures that aim to establish not only products but also the whole company as a platform.

Long before the two devices were conceptualized – around the year 2005 – Haier's visionary CEO Zhang Ruimin had already started to prepare the company's internal structures and processes for the agile development and product management needed in the digital era.

'The internal reorganization went in step with the fundamental reorientation of the whole Haier group towards embedding the whole range of our domestic appliances into the Internet of Things', Wo Yong recalls. To achieve this, the colossal organization was divided into hundreds of micro-enterprises as the most basic cells in a broad delivery platform of domestic appliances with hardly any hierarchy. In each of these units, swift innovation was pushed by direct communication and planning, by extremely decentralized decision making, and by direct two-way interaction between micro-enterprises within Haier and their end-user communities out there in daily life.

'No longer do successful companies compete through their brands. Instead, they compete through platforms – or, put another way, through linkages between independent enterprises, aligned via their interoperable technologies and their creative efforts', Zhang stated in a recent commentary piece reflecting on his company's repositioning towards the digital world.[13] Heeding that creed, Haier drastically changed its way of working to allow for maximum flexibility, creative freedom and consumer involvement in the management of Haier's product lifecycles.

The reformer teams gave the new model the Chinese name *rendanheyi*, where *ren* refers to the employees, *dan* means user value, and *heyi* indicates unity and an awareness of the whole system. 'The term *rendanheyi* suggests that employees can realize their own value during the process of creating value for users. This new model was intended to foster co-creation and win-win solutions for employees and customers', Zhang explains.

Three crucial changes to the set-up brought success

According to Zhang, the new approach to business is marked by three main characteristics and fundamental changes. First, the enterprise was transformed from a closed to an open system. This shift happened by the introduction of a network of self-governing micro-enterprises with, as Zhang puts it, 'free-flowing lateral communication between them and mutually creative connections with outside contributors.'[14] Second, the roles of employees are transformed from being executors of hierarchical top-down directions to self-motivated idea contributors, in many cases choosing or electing the leaders and members of their teams. In a crucial third move, the consumer purchasers of Haier's appliances are transformed, in the perspective of development and management teams, from traditional customers to lifetime users of products and services designed to solve their problems and enhance their user experience.

For the development of the smart wine cooler and the intelligent refrigerator, dedicated micro-enterprises were created, each with its own CEO. These executives brought in all the necessary talent and even created external entities to support the establishment of the product in the market. There are now 60 to 70 people working for the wine cooler micro-enterprise, for instance, says Wu Yong: 'It is a mix of people. Software engineers, of course, but also teams installing the coolers for the customers, application developers and operations staff.'

Haier's CEO puts the new multi-cellular set-up of his organization in a nutshell as follows:

'In effect, implementing the *rendanheyi* model meant tearing apart the walls of our enterprise and changing our structure into a collection of entrepreneurial ventures. The Haier platform now connects more than

2,000 microenterprises in various locations around the world. The leaders of each microenterprise have the power to make decisions, hire staff, and control distribution that would ordinarily accrue to the CEO of a company, not to a division leader. They can also manage the capital, recruiting external venture capital and conducting follow-up investment. They are, in effect, partners in their area of the enterprise. Only by this means can new opportunities be secured quickly; only when microenterprises are booming can a company the size of Haier maintain the passion and vitality of a pioneer.'

PART FOUR

Future product realities

13

Outlook 2030: How the reinvented product governs our lives – a crowd-sourced story of innovation in five takes

So far in this book we have focused on the big shifts and the evolutionary path required to reinvent products: their evolving path to become platforms, their development towards being a container for software and a carrier of services, their fundamental role change from being a provider of outputs to one of high-charged user experience outcomes, among other aspects.

All these shifts and developments will have no defined ending and we firmly expect them to run their course for many years to come. Given the super-fast progress of software and digital technology innovation, we wanted to end this book and hopefully extend its shelf life by providing a visionary perspective on how smart connected products will be embedded into businesses and society in 2030. Will quantum computing have become a household technology by then? What will AI be able to do? Will 4D printing be available to industrial businesses? Will blockchain technology have become a ubiquitous ingredient of future products and solutions?

To complement the more stringent analytical thoughts and prescriptive, practical business advice in the earlier chapters, this section takes a playful look at the potential evolution of smart connected products in a colourful, descriptive way. To develop this longer-term perspective, we leveraged several sources. First, we engaged Accenture's global employee base in a crowd-sourced innovation event over two days that generated over 2,000 inputs. Second, we engaged Accenture's Technology Labs and researchers to explore several specific future use cases. Finally, we leveraged our partner ecosystem for a variety of interviews that touched on the evolution of the market.

As a result, we have collated episodic scenes of day-to-day situations that try to anticipate how households, businesses, transport systems, cities, farms, shop floors – in short the B2C and B2B sphere – will most likely function in 12 years from now, given the rapid spread of smart connected products.

Take 1: Farming means high tech

Jack Monroe, 35, is a Wyoming-based farmer. The Monroe family grows sugar beets, dry beans, wheat and corn in the fifth generation. Like his granddad Walter, Jack still gets up early on a Wednesday in late August 2030, at 7:30 am. But unlike his ancestors, one of the first things to do is to start the home-schooling channel for his twin daughters, who take C++ coding lessons at the age of six, right after breakfast. Both love and have inherited dad's knack for everything tech.

Jack sits now on a reversed chair in front of two large screens in his 'command room', as he likes to refer to a central control cubicle for digital technology in his farmhouse. Moving two Xbox-style consoles jolts his workstation into action. Via camera Jack can now see the interior of the drone hangar he built three years ago to house the flying farm equipment he leases on a yearly basis from a local provider. Eight of his ten drones are lined up on their pads ready for action.

Sipping on a coffee, Jack analyses the data that has come in overnight. Analytics tools from a cloud provider have analysed the data and developed an optional flight pattern for his small drone fleet. The GPS flying coordinates are autonomously inputted into each of the waiting drones, where they are squared with weather and soil data provided by the drone maker.

Counting 2,300 acres, the Monroe farm is mid-sized by Wyoming standards. Three of the drones have overflown the land overnight identifying areas where weak plant sizes point to the need for fertilizer. Infrared imaging, speed-analysed by artificial intelligence tools, shows Jack where pointed watering should be administered and dark patches marked by the smart drone cameras show where a beetle outbreak needs to be contained by targeted spraying.

Jack has embedded advanced agricultural technology into the business since he took over. He passed the Remote Pilot Certificate required by the FAA for operating agricultural drones and goes on weekend courses on statistics and data analytics provided by the local community college to Wyoming wheat farmers. The need to feed the world's population has pushed high-precision agriculture to new levels and sophisticated digital farming equipment is now standard on over 80 per cent of all US farms.

Jack now comes in from the hangar where he has loaded the drones with water-based nitrogen plant nutrition and organic beetle repellent. With a few clicks the flying armada is cleared for take-off. A push of a button slides the hangar doors open and the drones swarm out into the fields to do their work. Meanwhile Jack turns to checking today's wheat prices online and booking a fully robotic autonomous harvester at his local 'agri outcome centre', as the wheat is ready to be cut in only a fortnight and hired equipment is usually difficult to come by in a busy harvesting season. The 'agri outcome' company charges Jack a fee per ton of wheat harvested, which aligns Jack's costs with his income and avoids the need for large capital equipment outlays.

The farmer has a framed gallery of faded photos sitting on the mantlepiece showing granddad Walter mounting a tractor at the Monroe Farm in 2019. Next to them sit the shiny trophies and medals Jack won for being among the 10 most productive farmers in his home state in 2029 and 2030.

Take 2: Private life is fully digitally curated

It is early morning in Kyoto, Japan. One of the country's popular home assistants, named Haiku, sits on a stylish nightstand next to the bed of Akari Suzuki, a 26-year-old primary school teacher.

The device starts playing back a serene wake-up melody, a cover version of the first song Akari learned from her grandmother when she was a child. The device's display shows 3 March 2030. Before playing the melody, it had analysed Akari's sleep cycle during the night based on sensors in her futon mattress and pillows and linked it with data from hundreds of previous nights to wake her up on this Tuesday morning at exactly 6:49 am, which was 10 minutes earlier than normal, but was the optimal time to wake, as she had finished an REM cycle and was in shallow sleep.

However, Haiku is much more than a personalized alarm clock. It backs up its owner as a fully fledged personal advocate platform. Having sat down for breakfast, Akari is now briefed by Haiku on her agenda for today. Akari asks a few questions on the agenda, and Haiku answers seamlessly using completely human-like full sentences in a conversational tone. One conflict has escaped her until now: a pupil's parent has asked for a phone call at 12:30 when Akari's teaching pauses at midday. She says 'rearrange by one hour' and the device does the rest, phoning the parent, coordinating diaries, and updating the meeting invite.

As Akari pours more green tea the AI-enabled device starts to make suggestions for dinner, taking into account that Akari is planning to go to the gym after work and is scheduled for a 760-calorie workout. Akari likes the suggestions but she overrides the device's plans by saying 'No', as she decided yesterday to join some colleagues for a sushi dinner after work. Akari asks Haiku to book a table for six people at the restaurant near her school. She gets an immediate confirmation and Haiku automatically sends an invite with the restaurant details to all attendees. Akari also welcomes Haiku's advice to put on a sweater as March days in Kyoto can still be chilly. The device has calculated a typical Tuesday morning time span between waking up and leaving home of exactly one hour and Akari can rely on Haiku making sure an autonomous vehicle is gliding up the driveway to her apartment block at 7:49 am. The autonomous vehicle takes a different route than usual today because of construction, but Akari still arrives at school well before the 8:30 am start.

Take 3: Ownership. Owner... what?

Zbigniev Lewandowski is a student reading business law at Krakow University, Poland. He rushes up his alma mater's imperial entrance hall staircase. He is currently pretty much under the gun as he will be trying to pass his master's degree in IP law next year. Inside the lecture halls and by crunching on many online tutorials he is trying to get his head around the complex legal structures entailed by data-sharing ecosystem alliances. In 2030 this is a young though rapidly growing legal field. Not least from his own day-to-day life, Zbigniev knows well that individual ownership of products has become almost entirely a concept of the past, since smart products have reached mass-market maturity. The law student owns practically nothing. For even the most mundane things like doing the washing, getting food, clothes and even housing, the 23-year-old Pole has subscribed to various lifecycle fulfilment packages that aggregate hundreds of services from B2B platform ecosystems, that own, or source, all the products he needs – from wedding day shoes to sunshades to falafel wraps or his university education material.

Businesses such as mortgages and car loans died out around the year 2025. And now, five years on, flats or vehicles, like practically all other products, are entirely hired and provided for the moment and exact time they are needed. Such services are clicked from the rich online offerings of established global outcome warehouses such as 'Outcomes 247' or 'Outsmart Direct', the latter being the jersey sponsor for Zbigniev's favourite football club Real Madrid.

Such comprehensive on-demand product lifestyles have been made possible by breakthroughs in materials science, which redefined product innovation and lifecycles. And technology advancement in clean energy and AI has made transportation cheap. It's what Zbigniev calls the 'Everything Economy', and it defines the daily life of a student whose parents were still used to pouring their life savings into things like an apartment or a decent car.

In his master's thesis the lawyer-in-waiting tries to unpick who legally owns the user data in a really tricky litigation case between a car-maker-turned-mobility-provider and a retail chain keen to use the inner-city mobility platform as its distribution channel. 'At least I have a job that no robot should be able to kill soon', he jokes. Data ownership and IP is one of the most protracted fields in company law as well as one of the few legal fields left that still requires human instead of AI representation in courts under EU regulation.

Take 4: Feel at home or at work on the road

At 6:24 pm a vehicle grinds to halt right in front of Anna Garcia. The senior nurse in her fifties is on the way to begin her night shift at a Buenos Aires hospital. Once the vehicle has stopped, the entire side panel of the car slides open to allow access to the comfortable seven-seater fully autonomous taxi van. Other passengers have already taken seats. Anna chucks her bag in and boards the vehicle, which is *en route* to the 'Hospital Naval'. It is a 30-minute ride through dense but highly organized inner-city traffic.

The city council of the Argentinian capital voted for fully orchestrated autonomous road traffic four years ago – no easy task in a sprawling urban agglomeration inhabited by 3.5 million people. But the system is now fully up and running and people like it. Around 1 million driverless cars are at any given time piloted through Buenos Aires by closely interacting on-board autopilots. Legacy car fleets have been fitted with the necessary technology to the cost of the public purse. As accident rates have come down close to zero, the council deemed the investment worthwhile as it can be recuperated by reduced spending in the city's hospitals.

Shared autonomous taxi vehicles, known as 'collectivos', which Anna has picked today for her trip to work, are extremely popular with commuter passengers as they offer compact but comfortable office environments to prepare for work shifts. For instance, extended reality eyewear helps Anna use the travel time to skim through patient medication files, work out shift plans for her team and watch short educational clips that will help her to take an exam to become a nurse anaesthetist next year.

Just before the vehicle pulls up before the hospital entrance, Anna instructs the on-board assistant to get in touch with her home assistant to check if the lights and cooker have been switched off safely. Within seconds she receives positive confirmation, as the vehicle's side slides open so that Anna can get out and new passengers in.

Take 5: The polite Frankenstein creatures we all love

'How do you take your coffee?' Zhang Wei turns around. He got tapped on the shoulder by a fluorescent-coloured service cobot taking orders for the morning break in Shanghai's largest manufacturing plant for autonomous vehicles. It is 9:30 am. 'Black, no sugar, a dash of milk and a whiff of cinnamon, as usual, Sir?' the fluent AI-voice assistant asks politely once it has identified Zhang Wei via face recognition cameras.

The 45-year-old Chinese manufacturing supervisor is known for his obsession with the accurate preparation of his coffee. 'We produce "customized lots of one" cars here, I expect the same from our coffee service', he says in work council meetings. Zhang Wei smiles at the cobot's precise recollection. He briefly nods, whereupon the cobot turns on its heels to elegantly glide on to the next human colleague who, like Zhang Wei, sits in a comfortably elevated command seat fitted with armrest controls to overview manufacturing floor.

The hall is brightly lit and the few human supervisors are only occasionally gesturing and talking to an army of intelligent self-organizing production robots and cobots whizzing around the hall fetching materials and assembling parts. The plant churns out 2,000 personalized autonomous cars per day. Around 3,000 humanoid all-rounders, made of steel, digital components and AI software, populate the floor supervised by Zhang Wei and only nine other human colleagues.

Closely interacting work groups of five to seven cobots are buzzing around the bare chassis skeletons inching forward on the nine assembly avenues in this hall. The AI-empowered machine brains, supported by high-resolution cameras and radar sensors, can decide on and do most assembly steps themselves – from gluing carbon fibre components to fitting personalized vehicle interiors such as HD screens for on-board entertainment to installing individually trained software autopilots.

Just every now and then, one of the cobots turns to their human supervisor for help, who then explains manufacturing steps by voice and hand gestures or gives extra information on special client needs for the machines to learn. On these brief educational sprints the robot gives a thumbs up to Zhang Wei and moves back to work, but not without remarking, 'I owe you a coffee, Zhang Wei... black, no sugar, a dash of milk and a whiff of cinnamon, was that?'

Scary? Not at all. Inspiring!

All the technologies working behind the described day-to-day scenes are available today, albeit some are more mature than others. In 2030, we believe, they will have evolved much further. And, as our (creatively guessed) scenarios can hopefully demonstrate, all of them will have moved on from today's pioneering beacon prototype status to accustomed mass-market realities transforming all our lives – as workers, consumers and businesses.

More productivity, more user experience, more personalization, more value and probably more leisure time will be the overall outcome for society of this broad but inevitable slide towards a world that is dominated by smart and connected products. Be it through autonomous cars, home assistant technology or the exploding market value, user data will command the future. Our communities and economies, the ways we will relate to them and to each other will be changing beyond recognition in a very short time span.

TAKEAWAYS AT A GLANCE

Chapter 1: The digital transformation of product making – happening faster than you think!

1 No industry is immune to the pervasiveness of digital. More than 75 per cent of industries are either at risk or are already significantly disrupted.

2 Digital is also rapidly overtaking hardware as the source of value in products. Companies need to follow the dual-drive innovation approach to digitally transform the core business while simultaneously creating a new breed of smart connected products.

3 There are six digital imperatives to navigate the change: transform the core; focus on experiences and outcomes; build or join ecosystems; innovate new business models; build a digital-ready workforce; and manage the wise pivots across your business.

Chapter 2: Trends driving the case for product reinvention

1 We are seeing the rapid rise of the outcome economy both in B2C and B2B.

2 In this new world, value and value creation are shifting from hardware to service and 'as a service'.

3 The age of mass customization is coming to an end; it will be replaced by the age of personal experience, use cases and context-specific services.

4 As a consequence, a complete re-architecturing of the product-making value chain and a transformation of the product development cycle are necessary.

Chapter 3: A radically new kind of product – adaptive | collaborative | proactive | responsible

1 A new product world is emerging in which products are becoming more intelligent and experience-rich. We can plot every product of this emerging new world against a new analytical tool that we call the Product Reinvention Grid.

2 Companies can manage to drive the increase of the IQ and the EQ. The magnitude of the management effort required to be successful can be estimated by combining the desired shift in IQ and EQ to calculate what we propose as the Product Reinvention Quotient (PRQ).

3 We have identified five big shifts. For most companies, a major transformation exploiting the five big shifts will have to be managed to succeed with their products in the new digitally driven world.

Chapter 4: Big Shift One: From features to experience

1 The basis of differentiation for a smart connected product is not traditional features and functions anymore, but holistic user experience.

2 Designing a compelling experience cannot be an afterthought. It must become an integral part of the product's value proposition and therefore be designed, engineered, monitored and refreshed as such.

3 As any product moves up the Experience Quotient (EQ) continuum, the experience becomes richer, broader and requires a robust ecosystem to power it.

Chapter 5: Big shift Two: From hardware to 'as a service'

1 Users are expecting to access as required rather than own a product, which leads to consuming the product as a service.

2 The software industry has proven that this shift can create enormous value. It is leading the way for the more hardware-centric industries such as automotive, industrial equipment or A&D, for example.

3 This transition is far from easy, and it requires major changes to operating models, product innovation processes, platforms and culture, as well as a complete re-architecting of the products.

Chapter 6: Big Shift Three: From product to platform

1 Platform business models are creating enormous market value.

2 Every product company must have a platform strategy, and determine whether to build their own platform or partner as well as what type of platform(s) models to participate in. Ignoring is not an option.

3 Many product companies will choose to partner with today's Internet giants who are platform leaders, but all need to understand the risk versus reward of their choices.

Chapter 7: Big Shift Four: From mechanical to artificial intelligence (AI)

1 AI technologies will rapidly be adopted in the majority of products, making them increasingly intelligent with abilities to sense, comprehend, act and learn.

2 Most product companies are still at the very early stages of incorporating AI technologies into their products. While close to 70 per cent of industrial companies believe that AI will transform their products and services, only 16 per cent have to date articulated a clear vision for AI and even fewer have a committed and funded roadmap.

3 Product companies need an increased sense of urgency to build AI capabilities and embed them in their product and experience roadmaps.

Chapter 8: Big Shift Five: From linear to agile engineering in the New

1 Traditional hardware product development is broken and will not work in a smart connected world. In the new world it is all about agility, iterations and experience.

2 A complete transformation of innovation using 'Engineering in the New' concepts and methods is required. Successful implementation can yield a 10X improvement in product development efficiency and effectiveness.

3 Unified data models and digital threading across the enterprise are required to enable next-generation products and as-a-service models.

Chapter 9: Seven pivotal capabilities for managing the reinvention of the product

1 To successfully reinvent your product, seven pivotal capabilities must be built.

2 While many of these new capabilities focus on the product development function, the shift to as-a-service business models affects virtually all processes and organizations.

3 In addition to building the new capabilities, a fundamental cultural and mindset change is required.

Chapter 10: The roadmap to success with living products and services

1 Virtually all product-making companies need to uplift the intelligence and experience quotients of their products, and to reinvent themselves into a living product or living service company.

2 To manage this transformation, a careful 'Rotation to the New' roadmap needs to be developed that includes digital transformation of today's core business to fund investment in the New. There are seven markers to help plot this roadmap.

3 Creation of a digital innovation factory is key for diffusing change and fuelling innovation as well as for attracting and retaining the skills needed.

4 Traditional organizational barriers need to be systematically torn down to encourage collaboration and increase agility.

GLOSSARY OF TERMS

2D/3D/4D	2-/3-/4-dimensional
3GPP	Third-Generation Partnership Project
4G/5G	Fourth/fifth generation of broadband cellular network technology
A&D	Aerospace & Defence
AI	Artificial Intelligence
ALM	Application Lifecycle Management
API	Application Programming Interface
AR	Augmented Reality
ASCD	Automatic Speed Control Device
ASIC	Application-Specific Integrated Circuit
AWS	Amazon Web Services
B2B	Business to Business
B2C	Business to Consumer
BIOS	Basic Input/Output System
BOM	Bill of Materials
CAD	Computer-Aided Design
CAM	Computer-Aided Manufacturing
CCD	Charge-Coupled Device
CEO	Chief Executive Officer
CIO	Chief Information Officer
CIP	Cockpit Intelligence Platform
CoE	Centre of Excellence
CoF	Cockpit of the Future
CPQ	Configuration, Pricing and Quoting
CRM	Customer Relationship Management
CSO	Chief Security Officer
CTO	Chief Technology Officer
CVP	Customer Value Proposition
DaaS	Device as a Service
DevOps	Development Operations
DevX	Develop for X

DFM	Design for Manufacturing
DfX	Design for Experience
DNA	Deoxyribonucleic Acid
Doc	Document
ECU	Electronic Control Unit
EQ	Experience Quotient
ERP	Enterprise Resource Planning
ExO	Exponential Organization
FAA	Federal Aviation Authority
F/O	Front Office
FOTA	Firmware Over the Air
FPGA	Field-Programmable Gate Array
FTE	Full-Time Equivalent
GB	Gigabyte
GPS	Global Positioning System
HD	High Definition
HE	Heavy Equipment
HMI	Human Machine Interface
HVAC	Heating, Ventilation & Air Conditioning
IDC	International Data Corporation
IEE	Industrial & Electrical Equipment
IIoT	Industrial Internet of Things
IML	Institute of Material Flow and Logistics
IoT	Internet of Things
IP	Intellectual Property
IPR	Intellectual Property Rights
IQ	Intelligence Quotient
IR	Infrared
IS	Information System
IT	Information Technology
KPI	Key Performance Indicator
LED	Light-Emitting Diode
LPWAN	Low Power Wide Area Network
LTE-CAT	Long-Term Evolution-Category
MBA	Master of Business Administration
MBOM	Manufacturing Bill of Materials
ML	Machine Learning

MPH	Miles Per Hour
MPN	Manufacturer's Part Number
MR	Mixed Reality
MRO	Maintenance, Repair and Overhaul
MVP	Minimum Viable Product
NB	Narrow Band
NFC	Near Field Communication
NLP	Natural Language Processing
OEM	Original Equipment Manufacturer
OES	Original Equipment Supplier
OS	Operating System
OT	Operating Technology
PC	Personal Computer
PDCA	Plan Do Check Adjust
PDM	Product Data Management
PLM	Product Lifecycle Management
PoS	Point of Sale
PRQ	Product Reinvention Quotient
R&D	Research and Development
RaaS	Robot as a Service
REM	Rapid Eye Movement
RF	Radio Frequency
RFID	Radio Frequency Identification
RMB	Renminbi
ROCE	Return on Capital Employed
RoHS	Restriction of Hazardous Substances
RPA	Robotic Process Automation
SaaS	Software as a Service
SDK	Solution Developer Kits
SKU	Stock Keeping Unit
SLA	Service Level Agreement
SQA	Software Quality Assurance
SRS	Supplemental Restraint System
SW	Software
TDD	Test Driven Development
TPU	Tensor Processing Unit
TTM	Time To Market

TV	Television
UI	User Interface
USB	Universal Serial Bus
UX	User Experience
VP	Vulnerability Patches
VR	Virtual Reality

ENDNOTES

Chapter 1

1 Paul, S (2017) Future perfect: the explosive impact of digitisation on the manufacturing industry, *Global Manufacturing* [online] http://www.manufacturingglobal.com/technology/future-perfect-explosive-impact-digitisation-manufacturing-industry [accessed 11 October 2018]

2 Roland Berger (2016) The Industrie 4.0 transition quantified, April [online] https://www.rolandberger.com/publications/publication_pdf/roland_berger_industry_40_20160609.pdf [accessed 11 October 2018]

3 World Economic Forum in collaboration with Accenture (2017) Digital Transformation Initiative: Unlocking $100 trillion for business and society from digital transformation, January [online] http://reports.weforum.org/digital-transformation/wp-content/blogs.dir/94/mp/files/pages/files/dti-executive-summary-website-version.pdf [accessed 11 October 2018]

4 Nicas, J (2018) Apple is worth $1,000,000,000,000. two decades ago, it was almost bankrupt, *New York Times*, 2 August [online] https://www.nytimes.com/2018/08/02/technology/apple-stock-1-trillion-market-cap.html?hp&action=click&pgtype=Homepage&clickSource=story-heading&module=first-column-region&region=top-news&WT.nav=top-news [accessed 11 October 2018]

5 Accenture

6 Afhsar, V (2015) Connected cars: every company will be a software company, *Huffington Post,* 16 May [online] https://www.huffingtonpost.com/vala-afshar/connected-cars-every-comp_b_7291144.html [accessed 11 October 2018]

7 Accenture

8 Accenture (2017) Connected Business Transformation [online] https://www.accenture.com/t20170202T140056__w__/za-en/_acnmedia/PDF-22/Accenture_Connected_Business_POV_FINAL_Online%20Feb1.pdf,%20accessed%20October%2022,%202018 [accessed 11 October 2018]

9 Accenture

10 Interview conducted by Accenture for this book

11 Accenture (2018) Disruption need not be an enigma, 26 February [online] https://www.accenture.com/gb-en/insight-leading-new-disruptability-index [accessed 11 October 2018]

12 Mills, M (2016) Drone disruption: the stakes, the players, and the opportunities, *Forbes*, 23 March [online] https://www.forbes.com/sites/markpmills/2016/03/23/drone-disruption-the-stakes-the-players-and-the-opportunities/#2b449be87d0b [accessed 11 October 2018]

13 Park, K (2017) Airbus is looking towards a future of pilotless planes, *Independent*, 22 November [online] https://www.independent.co.uk/news/business/news/airbus-pilotless-planes-self-flying-aircraft-passenger-flights-cto-paul-eremenko-a8068956.html [accessed 11 October 2018]

Chapter 2

1 Michelin (2014) Effifuel™ from Michelin® solutions delivers fuel savings, 9 September [online] https://www.michelin.com/eng/media-room/press-and-news/press-releases/Products-Services/EFFIFUEL-from-MICHELIN-R-solutions-Delivers-Fuel-Savings [accessed 11 October 2018]

2 Barlett, P (2016) Rolls-Royce negotiating with key clients on power-by-the-hour, *Seatrade Maritime News*, 8 September [online] http://www.seatrade-maritime.com/news/europe/rolls-royce-negotiating-with-key-clients-on-power-by-the-hour.html [accessed 11 October 2018]

3 Kaeser Compressors (nd) Sigma air utility [online] https://us.kaeser.com/services/compressed-air-as-utility-service/ [accessed 11 October 2018]

4 HP (nd) Managed print services [online] http://www8.hp.com/h20195/v2/GetPDF.aspx/4AA7-1042ENUS.pdf [accessed 11 October 2018]

5 Kurman, M and Lipson, H (2018) Why the rise of self-driving vehicles will actually increase car ownership, *Singularity Hub*, 14 February [online] https://singularityhub.com/2018/02/14/why-the-rise-of-self-driving-vehicles-will-actually-increase-car-ownership/#sm.00t49dd81039fhd11pc1u7ns1xgqe [accessed 11 October 2018]

6 RethinkX (2017) New report: due to major transportation disruption, 95% of U.S. car miles will be traveled in self-driving, electric, shared vehicles by 2030 [online] https://www.rethinkx.com/press-release/2017/5/3/new-report-due-to-major-transportation-disruption-95-of-us-car-miles-will-be-traveled-in-self-driving-electric-shared-vehicles-by-2030 [accessed 11 October 2018]

7 Intel (2017) Intel predicts autonomous driving will spur new 'passenger economy' worth us$7 trillion, 6 January [online] https://www.intc.com/investor-relations/investor-education-and-news/investor-news/press-release-details/2017/Intel-Predicts-Autonomous-Driving-Will-Spur-New-Passenger-Economy-Worth-US7-Trillion/default.aspx [accessed 11 October 2018]

8 Accenture

9 Rossana Ricco Rodgers, Group VP, Head of AIM Product Management at ABB. 'This is my personal opinion based on my two decades of work experience in deploying digital products and does not represent in any way the ABB official statement in this matter.'

10 Markman, J (2017) Netflix knows what you want ... before you do, *Forbes*, 9 June [online] https://www.forbes.com/sites/jonmarkman/2017/06/09/netflix-knows-what-you-want-before-you-do/#30a4eaca52b8 [accessed 11 October 2018]

11 Accenture

12 Haier (2017) Haier Rendanheyi Management Model in Stanford University, 18 March [online] http://www.haier.net/en/about_haier/news/201703/t20170328_345989.shtml [accessed 11 October 2018]

13 Forsblom, N (2015) Were you aware of all these sensors in your smartphone? *Adtile*, 12 November [online] https://blog.adtile.me/2015/11/12/were-you-aware-of-all-these-sensors-in-your-smartphone/ [accessed 11 October 2018]

14 Accenture (2018) Accenture to demonstrate multi-passenger, in-vehicle Amazon Alexa Voice Service at CES 2018 in Las Vegas, 10 January [online] https://newsroom.accenture.com/news/accenture-to-demonstrate-multi-passenger-in-vehicle-amazon-alexa-voice-service-at-ces-2018-in-las-vegas.htm [accessed 11 October 2018]

Chapter 3

1 Fisher, T (2018) How are 4G and 5G so different? *Lifewire*, 31 October [online] https://www.lifewire.com/5g-vs-4g-4156322 [accessed 12 October 2018]

2 Forsblom, N (2015) Were you aware of all these sensors in your smartphone? *adtile*, 12 November [online] https://blog.adtile. me/2015/11/12/were-you-aware-of-all-these-sensors-in-your-smartphone/ [accessed 12 October 2018]

3 Rapolu, B (2016) Internet of aircraft things: an industry set to be transformed, *Aviation Week*, 18 January [online] http://aviationweek.com/ connected-aerospace/internet-aircraft-things-industry-set-be-transformed [accessed 12 October 2018]

4 Airbus (2018) Airbus' open aviation data platform Skywise continues to gain market traction, 7 February [online] https://www.airbus.com/ newsroom/press-releases/en/2018/02/airbus--open-aviation-data-platform-skywise-continues-to-gain-ma.html [accessed 12 October 2018]

5 Automotive Sensors (2017) 2017 Expo [online] http://www. automotivesensors2017.com/ [accessed 12 October 2018]

6 Trafton, A (2017) Flexible sensors can detect movement in GI tract, *MIT News*, 10 October [online] http://news.mit.edu/2017/flexible-sensors-can-detect-movement-gi-tract-1010 [accessed 12 October 2018]

7 Kapurwan, A (2014) Sensors used in washing machine and their functioning, *Prezi*, 28 April [online] https://prezi.com/ieqcgm5dde-f/ sensors-used-in-washing-machine-and-their-functioning/ [accessed 12 October 2018]

8 Routley, N (2018) How the computing power in a smartphone compares to supercomputers past and present, *Business Insider*, 7 November [online] https://www.businessinsider.de/infographic-how-computing-power-has-changed-over-time-2017-11?r=US&IR=T [accessed 12 October 2018]

9 Marr, B (2017) The amazing ways Coca Cola uses artificial intelligence and big data to drive success, *Forbes*, 18 September [online] https://www. forbes.com/sites/bernardmarr/2017/09/18/the-amazing-ways-coca-cola-uses-artificial-intelligence-ai-and-big-data-to-drive-success/#620fb80278d2 [accessed 12 October 2018]

10 Vincent, J (2017) John Deere is buying an AI startup to help teach its tractors how to farm, *The Verge*, 17 September [online] https://www.theverge.com/2017/9/7/16267962/automated-farming-john-deere-buys-blue-river-technology [accessed 12 October 2018]

11 Sincavage, D (nd) How artificial intelligence will change decision-making for businesses, *Tenfold* [online] https://www.tenfold.com/business/artificial-intelligence-business-decisions [accessed 12 October 2018]

12 Continuum Innovation (nd) Haier: Smart window refrigerator [online] https://www.continuuminnovation.com/en/what-we-do/case-studies/smart-window-refrigerator [accessed 12 October 2018]

13 Accenture

14 Interview conducted by Accenture for this book

15 Accenture

16 Tung, L (2018) Elon Musk: Tesla Autopilot gets full self-driving features in August update, *ZDnet*, 12 June [online] https://www.zdnet.com/article/elon-musk-tesla-autopilot-gets-full-self-driving-features-in-august-update/ [accessed 12 October 2018]

17 Interview conducted by Accenture for this book

18 Nest (nd) Learn how to control your Nest products with Amazon Alexa [online] https://nest.com/support/article/Nest-and-Amazon-Alexa [accessed 12 October 2018]

19 Pacific BMW (nd) How gesture control in BMW 7 series works [online] http://www.pacificbmw.com/blog/how-gesture-control-in-bmw-7-series-works/ [accessed 12 October 2018]

20 Faurecia (nd) Faurecias sprachgesteuertes cockpit der zukunft gewinnt Industriepreis 2018 [online] https://newsroom.faurecia.de/news/faurecias-sprachgesteuertes-cockpit-der-zukunft-gewinnt-industriepreis-2018-b756-0a54a.html#IoXKTp58cRThO49q.99https://newsroom.faurecia.de/news/faurecias-sprachgesteuertes-cockpit-der-zukunft-gewinnt-industriepreis-2018-b756-0a54a.html [accessed 12 October 2018]

Chapter 4

1 Ward, M (2016) Omo creates smart clothes peg in an effort to make washing less of a chore, *Mumbrella*, 22 April [online] https://mumbrella.com.au/omo-smart-peg-j-walter-thompson-peggy-361828 [accessed 12 October 2018]

2 Wright, I (2017) Airbus uses smart glasses to improve manufacturing efficiency, *Engineering.com*, 28 March [online] https://www.engineering.com/AdvancedManufacturing/ArticleID/14634/Airbus-Uses-Smart-Glasses-to-Improve-Manufacturing-Efficiency.aspx [accessed 12 October 2018]

3 Softbank Robotics (nd) Pepper [online] https://www.softbankrobotics.com/emea/en/pepper [accessed 12 October 2018]

4 Ibid

5 Sumagaysay, L (2018) Sony's aibo robotic dog can sit, fetch and learn what its owner like, *Mercury News*, 14 September [online] https://www.mercurynews.com/2018/09/14/sonys-aibo-robotic-dog-can-sit-fetch-and-learn-what-its-owner-likes/ [accessed 12 October 2018]

6 Biesse Group (2018) Make, 16 May [online] https://issuu.com/biessegroup/docs/biessegroup_make_06_lr [accessed 12 October 2018]

7 Accenture (2018) Get personal: drive profitable growth as an intelligent B2B enterprise [online] https://www.accenture.com/t20180611T064843Z__w__/in-en/_acnmedia/PDF-58/Accenture-Drive-B2B-Sales-Growth.pdf#zoom=50 [accessed 12 October 2018]

8 Interview conducted by Accenture for this book

9 Wollan, R, Jacobson, T and Honts, R (2016) Digital disconnect in customer engagement, *Accenture* [online] https://www.accenture.com/mx-es/insight-digital-disconnect-customer-engagement [accessed 12 October 2018]

10 Ibid

11 Schaeffer, E (2017) *Industry X.0*, Kogan Page, pp 136–37, Figures 5.4 and 5.5

12 World Economic Forum in collaboration with Accenture (2017) Digital transformation initiative: mining and metals industry, January [online] http://reports.weforum.org/digital-transformation/wp-content/blogs.dir/94/mp/files/pages/files/wef-dti-mining-and-metals-white-paper.pdf [accessed 12 October 2018]

13 Reisinger, D (2017) Ikea's new smart lights will power up from Apple, Amazon, and Google devices, *Fortune*, 1 November [online] http://fortune. com/2017/11/01/ikea-smart-lights-alexa-apple/ [accessed 12 October 2018]

14 Shook, E and Knickrehm, M (2018) Reworking the revolution, *Accenture* [online] https://www.accenture.com/t20180613T062119Z__w__/us-en/_ acnmedia/PDF-69/Accenture-Reworking-the-Revolution-Jan-2018-POV. pdf#zoom=50 [accessed 12 October 2018]

15 Ibid

16 Forrester Consulting (2016) Expectation vs experience: the good, the bad, the opportunity, *Accenture* [online] https://www.accenture.com/ t20160825T041338Z__w__/us-en/_acnmedia/PDF-23/Accenture- Expectations-Vs-Experience-Infographic-June-2016.pdfla=en [accessed 12 October 2018]

17 Accenture

18 Interview conducted by Accenture for this book

19 Ibid

20 Ibid

Chapter 5

1 Accenture

Chapter 6

1 Trefis and Great Speculations (2018) As a rare profitable unicorn, Airbnb appears to be worth at least $38 billion, *Forbes*, 11 May [online] https:// www.forbes.com/sites/greatspeculations/2018/05/11/as-a-rare-profitable- unicorn-airbnb-appears-to-be-worth-at-least-38-billion/#211df9722741 [accessed 12 October 2018]

2 Randewich, N (2018) Stock market value of Netflix eclipses Disney for the first time, *Reuters*, 24 May [online] https://www.reuters.com/article/us-netflix-stocks-marketcap/stock-market-value-of-netflix-eclipses-disney-for-first-time-idUSKCN1IP39C [accessed 12 October 2018]

3 Accenture

4 Interview conducted by Accenture for this book

5 Smith, Craig (2018) 15 interesting Waze statistics and facts, *DMR*, 27 October [online] https://expandedramblings.com/index.php/waze-statistics-facts/ [accessed 12 October 2018]

6 Holodny, E (2017) Equifax CEO steps down after massive data breach, *Business Insider*, 26 September [online] https://www.businessinsider.de/equifax-ceo-out-2017-9?r=US&IR=T [accessed 12 October 2018]

7 Finkle, J (2014) Hackers raid eBay in historic breach, access 145 million records, *Reuters*, 21 May [online] https://uk.reuters.com/article/uk-ebay-password/hackers-raid-ebay-in-historic-breach-access-145-million-records-idUKKBN0E10ZL20140522 [accessed 12 October 2018]

8 Shubber, K (2018) Yahoo's $35m fine sends a message, *FT*, 12 July [online] https://www.ft.com/content/4c0932f0-6d8a-11e8-8863-a9bb262c5f53 [accessed 12 October 2018]

9 Interview conducted by Accenture for this book

10 Winton, N (2017) British vacuum maker Dyson plans electric car assault, *Forbes*, 26 September [online] https://www.forbes.com/sites/neilwinton/2017/09/26/dyson-british-vacuum-cleaner-plans-electric-car-assault-with-2-7-billion-plan/ [accessed 12 October 2018]

11 Interview conducted by Accenture for this book

12 Accenture (2016) Platform economy: technology-driven business model innovation from the outside in [online] https://www.accenture.com/t00010101T000000Z__w__/gb-en/_acnmedia/Accenture/Omobono/TechnologyVision/pdf/Platform-Economy-Technology-Vision-2016.pdfla=en-GB#zoom=50 [accessed 12 October 2018]

13 Lunden, I (2018) App Store hits 20M registered developers and $100B in revenues, 500M visitors per week [online] https://techcrunch.com/2018/06/04/app-store-hits-20m-registered-developers-at-100b-in-revenues-500m-visitors-per-week/ [accessed 6 December 2018]

14 Leswing, K (2018), Apple just shared some staggering statistics about how well the App Store is doing, Business Insider, 4 January [online] https://www. businessinsider.de/apple-app-store-statistics-2017-2018-1?r=US&IR=T [accessed 6 December 2018]

15 Rao, L (2016), Apple's App Store Has Achieved 140 Billion Downloads, Fortune, 7 September [online] http://fortune.com/2016/09/07/apple-app-downloads/ [accessed 6 December 2018]

16 Leswing, K (2018), Apple just shared some staggering statistics about how well the App Store is doing, Business Insider, 4 January [online] https:// www.businessinsider.de/apple-app-store-statistics-2017-2018-1?r=US&IR=T [accessed 6 December 2018]

17 Schneider Electric (2018) Next generation of ExoStructure power [online] https://www.schneider-electric.com/en/about-us/press/news/corporate-2018/ecostruxure-power.jsp [accessed 12 October 2018]

18 Faurecia (2018) Capital markets day Faurecia transformation, 15 May [online] http://www.faurecia.com/sites/groupe/files/pages/20180515-investor-day-presentation-en_0.pdf [accessed 12 October 2018]

19 Murray, S (2015), IDC Predicts the Emergence of "the DX Economy" in a Critical Period of Widespread Digital Transformation and Massive Scale Up of 3rd Platform Technologies in Every Industry, Business Wire, 4 November [online] https://www.businesswire.com/news/home/20151104005180/en/IDC-Predicts-Emergence-DX-Economy-Critical-Period [accessed 7 December 2018]

Chapter 7

1 Daugherty, P and Wilson, J (2018) *Human+Machine: Reimagining work in the age of AI*, Harvard Business Review Press

2 Wilson, J, Alter, A and Sachdev, S (2016) Business processes are learning to hack themselves, *Harvard Business Review*, 27 June [online] https://hbr.org/2016/06/business-processes-are-learning-to-hack-themselves [accessed 12 October 2018]

3 Knight, W (2016) This factory robot learns a new job overnight, *MIT Technology Review*, 18 March [online] https://www.technologyreview.com/s/601045/this-factory-robot-learns-a-new-job-overnight/ [accessed 12 October 2018]

4 PPD (2018) PPD names Colin Hill, CEO of GNS Healthcare, to its board of directors, 3 January https://www.ppdi.com/News-And-Events/News/2018/PPD-Names-Colin-Hill-CEO-of-GNS-Healthcare-to-its-Board-of-Directors [accessed 12 October 2018]

5 Gartner (2018) Gartner forecasts worldwide public cloud revenue to grow 21.4 percent in 2018, 12 April [online] https://www.gartner.com/newsroom/id/3871416 [accessed 12 October 2018]

6 Business Wire (2017) $92.48 billion cloud storage market – forecasts from 2017 to 2022 – research and markets [online] https://www.businesswire.com/news/home/20170614005856/en/92.48-Billion-Cloud-Storage-Market–Forecasts [accessed 12 October 2018]

7 Ismail, K (2017) The value of data: forecast to grow 10-fold by 2025, Information Age, 5 April [online] https://www.information-age.com/data-forecast-grow-10-fold-2025-123465538/ [accessed 6 December 2018]

8 Pestanes, P and Gautier, B (2017) The rise of intelligent voice assistants, *Wavestone* [online] https://www.wavestone.com/app/uploads/2017/09/Assistants-vocaux-ang-02-.pdf [accessed 12 October 2018]

9 Dormehl, L (2018) Today in Apple history: Siri makes its public debut on iPhone 4s, *Cult of Mac*, 4 October [online] https://www.cultofmac.com/447783/today-in-apple-history-siri-makes-its-public-debut-on-iphone-4s/ [accessed 12 October 2018]

10 Tillman, M and Grabham, D (2018) What is Google Assistant and what can it do? *Pocket Lint*, 9 October [online] https://www.pocket-lint.com/apps/news/google/137722-what-is-google-assistant-how-does-it-work-and-which-devices-offer-it [accessed 12 October 2018]

11 Callaham, J (2018) Amazon Echo is now available for everyone to buy for $179.99, shipments start on July 14, *Android Central*, 23 June [online] https://www.androidcentral.com/amazon-echo-now-available-everyone-buy-17999-shipments-start-july-14 [accessed 12 October 2018]

12 Lopez, N (2016) Google Home finally has a price and release date, *The Next Web*, 4 October [online] https://thenextweb.com/google/2016/10/04/google-shows-off-home-can-take-amazons-echo/ [accessed 12 October 2018]

13 Macrumours (2018) Homepod: Apple's Siri-based speaker, available now [online] https://www.macrumors.com/roundup/homepod/ [accessed 12 October 2018]

14 Kinsella, B (2018) 56 million smart speaker sales in 2018 says Canalys, *Voicebot*, 7 January [online] https://voicebot.ai/2018/01/07/56-million-smart-speaker-sales-2018-says-canalys/ [accessed 12 October 2018]

15 Kastrenakes, J (2018) Google Assistant will soon detect what language you're speaking in, *The Verge*, 23 February [online] https://www.theverge.com/2018/2/23/17041920/google-assistant-languages-multilingual-detection [accessed 12 October 2018]

16 Leswing, K (2018) Google Assistant tops Apple's Siri and Amazon's Alexa in head-to-head intelligence test, *Business Insider*, 4 August [online] https://www.businessinsider.de/google-assistant-vs-apple-siri-amazon-alexa-comparison-2018-7?r=US&IR=T [accessed 12 October 2018]

17 Accenture (2018) Time to navigate the super myway [online] https://www.accenture.com/us-en/_acnmedia/PDF-69/Accenture-2018-Digital-Consumer-Survey-Findings.pdf [accessed 12 October 2018]

18 Seung, H L et al (2017) Implantable batteryless device for on-demand and pulsatile insulin administration, *Nature*, 13 April [online] https://www.nature.com/articles/ncomms15032 [accessed 12 October 2018]

19 Interview conducted by Accenture for this book

20 Accenture (2018) Turning possibility into productivity [online] https://www.accenture.com/t00010101T000000Z__w__/gb-en/_acnmedia/PDF-76/Accenture-IndustryX0-AI-products.pdf [accessed 12 October 2018]

21 Ibid

22 3M (nd) Natural Language Processing. Applied to make use of unseen data [online] https://www.3m.com/3M/en_US/health-information-systems-us/providers/natural-language-processing/ [accessed 12 October 2018]

23 Agri Expo (nd) Bosch: Mehrzweck-agrarroboter/für unkraut [online] http://www.agriexpo.online/de/prod/bosch-deepfield-robotics/product-168586-1199.html [accessed 12 October 2018]

24 Blue River Technology (nd) About us [online] http://about.bluerivertechnology.com/ [accessed 12 October 2018]

25 Accenture Research

Chapter 8

1 Porter, M and Happleman, J (2015) How smart, connected products are transforming companies, *Harvard Business Review*, October [online] https://hbr.org/2015/10/how-smart-connected-products-are-transforming-companies [accessed 12 October 2018]

2 Interview conducted by Accenture for this book

3 Schrage, M (2016) Fast, iterative "virtual research centers" are edging out traditional approaches to R&D, *MIT Sloan Management Review*, 11 May [online] https://sloanreview.mit.edu/article/rd-meet-es-experiment-scale/ [accessed 12 October 2018]

4 Krawiec, T (nd) The Amazon Recommendations secret to selling more online, *Rejoiner* [online] http://rejoiner.com/resources/amazon-recommendations-secret-selling-online/ [accessed 12 October 2018]

5 Hern, A (2018) The two-pizza rule and the secret of Amazon's success, *Guardian*, 24 April [online] https://www.theguardian.com/technology/2018/apr/24/the-two-pizza-rule-and-the-secret-of-amazons-success [accessed 12 October 2018]

6 Schrage, M (2016) Fast, iterative "virtual research centers" are edging out traditional approaches to R&D, *MIT Sloan Management Review*, 11 May [online] https://sloanreview.mit.edu/article/rd-meet-es-experiment-scale/ [accessed 12 October 2018]

7 Accenture

8 Porter, M and Happleman, J (2015) How smart, connected products are transforming companies, *Harvard Business Review*, October [online] https://hbr.org/2015/10/how-smart-connected-products-are-transforming-companies [accessed 12 October 2018]

9 Ibid

10 Interview conducted by Accenture for this book

11 Accenture

12 Ibid

13 Accenture (2017) The digital thread imperative [online] https://www.accenture.com/t20171211T045641Z__w__/us-en/_acnmedia/PDF-67/Accenture-Digital-Thread-Aerospace-And-Defense.pdf [accessed 12 October 2018]

Chapter 9

1 Fanguy, W (2018) How Netflix designs with flexibility, *Inside Design*, 28 March [online] https://www.invisionapp.com/blog/netflix-designs-flexibility/ [accessed 12 October 2018]

2 Accenture (2017) Accenture helps machinery manufacturing company Biesse harness Industry X.0 for connected customer services, 6 November [online] https://newsroom.accenture.com/news/accenture-helps-machinery-manufacturing-company-biesse-group-harness-industry-x-0-for-connected-customer-services.htm [accessed 12 October 2018]

3 Interview conducted by Accenture for this book

4 Fanguy, W (2018) How Netflix designs with flexibility, *Inside Design*, 28 March [online] https://www.invisionapp.com/blog/netflix-designs-flexibility/ [accessed 12 October 2018]

5 Agriculture.com community page [online] https://community.agriculture.com/ [accessed 12 October 2018]

6 Stanton, R (2014) How Schneider Electric is learning from startups, *Schneider*, 3 June [online] https://blog.schneider-electric.com/education-research/2014/06/03/schneider-electric-learning-startups/ [accessed 12 October 2018]

7 Interview conducted by Accenture for this book

8 Davenport, T, Kudbya, S and Paul, S (2017) IoT and developing analytics-based data products, *MIT Sloan Management Review*, 9 January [online] https://sloanreview.mit.edu/article/iot-and-developing-analytics-based-data-products/ [accessed 12 October 2018]

9 To learn more about ecosystems please refer to Chapter 9 'Zoom in: How to Make the Most of Platforms and Ecosystems' in the following book: Schaeffer, E (2017) *Industry X.0,* Kogan Page

10 Interview conducted by Accenture for this book

11 Ibid

Chapter 10

1 Accenture (2018) Turning possibility into productivity [online] https://www.accenture.com/t00010101T000000Z__w__/gb-en/_acnmedia/PDF-76/Accenture-IndustryX0-AI-products.pdf [accessed 12 October 2018]

2 Interview conducted by Accenture for this book

3 Ibid

4 Ibid

5 Walker, R (2017) How Adobe got its customers hooked on subscriptions, *Bloomberg Businessweek*, 8 June [online] https://www.bloomberg.com/news/articles/2017-06-08/how-adobe-got-its-customers-hooked-on-subscriptions [accessed 12 October 2018]

6 Accenture

7 Zhang, R (2018) Why Haier is reorganizing itself around the Internet of Things, *Strategy+Business*, 26 February [online] https://www.strategy-business.com/article/Why-Haier-Is-Reorganizing-Itself-around-the-Internet-of-Things?gko=895fe [accessed 12 October 2018]

8 Sparkes, M (2018) Tesla software update: did your car just get faster? *Telegraph*, 30 January [online] https://www.telegraph.co.uk/technology/news/11378880/Tesla-software-update-did-your-car-just-get-faster.html [accessed 12 October 2018]

9 Hawkins, A (2018) Tesla will start rolling out its 'full self-driving' package in August, Elon Musk says, *The Verge*, 11 June [online] https://www.theverge.com/2018/6/11/17449076/tesla-autopilot-full-self-driving-elon-musk [accessed 12 October 2018]

10 Interview conducted by Accenture for this book

11 Accenture (2017) News Release: Schneider Electric and Accenture build a digital services factory to speed development of industrial IoT solutions and services, 26 April [online] https://newsroom.accenture.com/news/schneider-electric-and-accenture-build-a-digital-services-factory-to-speed-development-of-industrial-iot-solutions-and-services.htm [accessed 12 October 2018]

12 Accenture (nd) Charging into the new: Smart processes bring smart new digital services for Schneider Electric [online] https://www.accenture.com/in-en/success-schneider-electric-digital-services-factory [accessed 12 October 2018]

13 Ibid

14 Interview conducted by Accenture for this book

15 Ibid

16 Accenture (2017) Accenture helps machinery manufacturing company Biesse harness Industry X.0 for connected customer services, 6 November [online] https://newsroom.accenture.com/news/accenture-helps-machinery-manufacturing-company-biesse-group-harness-industry-x-0-for-connected-customer-services.htm [accessed 12 October 2018]

17 Schaeffer, E et al (2018) AI turns ordinary products into industry game-changers, *Accenture*, 20 April [online] https://www.accenture.com/us-en/insights/industry-x-0/ai-transforms-products [accessed 12 October 2018]

Chapter 12

1 Lienert, P (2016) France's Faurecia recasts interiors for self-driving cars, *Reuters*, 25 August [online] https://www.reuters.com/article/us-faurecia-outlook/frances-faurecia-recasts-interiors-for-self-driving-cars-idUSKCN1102CU [accessed 12 October 2018]

2 Market Screener (nd) Faurecia [online] https://www.marketscreener.com/FAURECIA-4647/company/ [accessed 12 October 2018]

3 Faurecia (2018) Capital markets day Faurecia transformation, 15 May [online] http://www.faurecia.com/sites/groupe/files/pages/20180515-investor-day-presentation-en_0.pdf [accessed 18 October 2018]

4 Signify (2018) Investor presentation [online] https://www.signify.com/static/events/signify-latest-investor-presentation.pdf [accessed 12 October 2018]

5 Zhang, R (2018) Why Haier is reorganizing itself around the Internet of Things, *Strategy+Business*, 26 February [online] https://www.strategy-business.com/article/Why-Haier-Is-Reorganizing-Itself-around-the-Internet-of-Things?gko=895fe [accessed 12 October 2018]

6 Haier (2018) Haier rebrands as global leading smart home solution platform for IoT era to showcase new smart home solutions ahead of 2018 AWE, *PR Newswire*, 7 March [online] https://www.prnewswire.com/news-releases/haier-rebrands-as-global-leading-smart-home-solution-platform-for-iot-era-to-showcase-new-smart-home-solutions-ahead-of-2018-awe-300609618.html [accessed 12 October 2018]

7 Ibid

8 Runckel, C (2012) Wine industry in China, *Business in Asia* [online] http://www.business-in-asia.com/china/china_wine.html [accessed 12 October 2018]

9 Chen, M (2018) By 2021, China to replace UK as No 2 on global wine market list, *China Daily*, 26 March [online] http://www.chinadaily.com.cn/a/201803/26/WS5ab8347aa3105cdcf651422f.html [accessed 12 October 2018]

10 Willsher, K (2014) China becomes biggest market for red wine, with 1.86bn bottles sold in 2013, *Guardian*, 29 January [online] https://www.theguardian.com/world/2014/jan/29/china-appetite-red-wine-market-boom [accessed 12 October 2018]

11 RFID technology is a universally adopted radio frequency identification. The chip is commonly used and uses electromagnetic fields to automatically identify and track tags attached to objects. There is no problem in identification as long as there is data onboarding. For now, the agency of drinks and beverages is responsible for tagging the chip on the bottle when it enters China Customs. In further cooperation, the chateau overseas may tag the chip directly before exporting it to China.

12 Willsher, K (2014) China becomes biggest market for red wine, with 1.86bn bottles sold in 2013, *Guardian*, 29 January [online] https://www.theguardian.com/world/2014/jan/29/china-appetite-red-wine-market-boom [accessed 12 October 2018]

13 Zhang, R (2018) Why Haier is reorganizing itself around the Internet of Things, *Strategy+Business*, 26 February [online] https://www.strategy-business.com/article/Why-Haier-Is-Reorganizing-Itself-around-the-Internet-of-Things?gko=895fe [accessed 12 October 2018]

14 Ibid

INDEX

The index is filed in alphabetical, word-by-word order. Numbers and acronyms within main headings are filed as spelt out. Page locators in italics denote information contained within a figure or table.